Brain Function and Responsiveness in Disorders of Consciousness

Martin M. Monti • Walter G. Sannita
Editors

Brain Function and Responsiveness in Disorders of Consciousness

Springer

Editors
Martin M. Monti
Department of Psychology
and Neurosurgery
University of California Los Angeles
Los Angeles, CA
USA

Walter G. Sannita
Department of Neuroscience
Ophthalmology and Genetics
University of Genova
Genova
Italy

Department of Psychiatry
State University of New York
at Stony Brook
Stony Brook, NY
USA

ISBN 978-3-319-21424-5 ISBN 978-3-319-21425-2 (eBook)
DOI 10.1007/978-3-319-21425-2

Library of Congress Control Number: 2015957244

Springer Cham Heidelberg New York Dordrecht London

Printed on acid-free paper

Springer International Publishing AG Switzerland is part of Springer Science+Business Media (www.springer.com)

Preface

Over the past 20 years, functional neuroimaging has revealed that even the severely injured brain suffering from a disorder of consciousness can retain several aspects of relatively high-level functions, including sensory and linguistic processing and learning dynamics. In 2006, a provocative report presented the case of a subject in vegetative state who appeared able to voluntarily engage in mental tasks as revealed by her pattern of brain activations.

As widely reported as these neuroimaging studies have been, the interpretation of regional brain activations in the vegetative state remains highly debated. These observations nevertheless challenge the current definitions and our understanding of both responsiveness and consciousness after severe brain injury, with an impact in the clinical decision-making process. It is unclear the extent to which regional brain activations can be considered equivalent to behavioral responses in indicating consciousness. The very clinical standards by which patients surviving severe brain injury are judged as being conscious or unconscious have been called into question. In this respect, scientific research has somehow added a conceptual and professional burden to the clinician by introducing novel criteria of evaluation not yet fully integrated in the current nosography of disorders of consciousness, which is now undergoing a tacit, but not uncontroversial, revision.

This multiauthored book is aimed at reviewing the scientific evidence available to date and the current different points of view in the field. Its task is to provide neuroscientists and clinicians (as well as ethicists, philosophers, and lawyers) with tools to incorporate the recent scientific advances in the understanding of disorders of consciousness.

Los Angeles, CA, USA
Genova, Italy
Stony Brook, NY, USA

Martin M. Monti, PhD
Walter G. Sannita, MD

Contents

Contributors

Enrico Amico Coma Science Group, Neurology Department, Cyclotron Research Center, University of Liège, Liège, Belgium

R. David Andrew Department of Biomedical & Molecular Sciences, Queen's University, Kingston, ON, Canada

Jitka Annen Coma Science Group, Neurology Department, Cyclotron Research Center, University of Liège, Liège, Belgium

Georgios Antonopoulos Coma Science Group, Neurology Department, Cyclotron Research Center, University of Liège, Liège, Belgium

James L. Bernat Department of Neurology, Geisel School of Medicine at Dartmouth, Hanover, NH, USA

Heinrich Binder Department für Klinische Medizin und Präventionsmedizin, Neurological Center, Karl Landsteiner Institut für Neurorehabilitation und Raumfahrtneurologie, Vienna, Austria

Neurological Centre, Otto Wagner Hospital, Vienna, Austria

Mélanie Boly Coma Science Group, Cyclotron Research Center and Neurology Department, University and University Hospital of Liège, Liège, Belgium

Daniel Bor Sackler Centre for Consciousness Science, University of Sussex, Brighton, UK

Department of Informatics, University of Sussex, Brighton, UK

Silvia Casarotto Department of Biomedical and Clinical Sciences Luigi Sacco, University of Milan, Milan, Italy

Carlo Cavaliere Coma Science Group, Neurology Department, Cyclotron Research Center, University of Liège, Liège, Belgium

Marianna Cavinato Fondazione Ospedale San Camillo, I.R.C.C.S., Venezia, Italy

Gastone G. Celesia Department of Neurology, Loyola University of Chicago, Maywood, IL, USA

Chicago Council on Science and Technology, Chicago, IL, USA

Damian Cruse The Brain and Mind Institute, University of Western Ontario, London, ON, Canada

Carol Di Perri Coma Science Group, Nceurology Department, Cyclotron Research Center, University of Liège, Liège, Belgium

Olivia Gosseries Coma Science Group, Cyclotron Research Center and Neurology Department, University and University Hospital of Liège, Liège, Belgium

Lizette Heine Coma Science Group, Cyclotron Research Centre, University of Liège, Liège, Belgium

L. Syd M. Johnson Department of Humanities, Michigan Technological University, Houghton, MI, USA

Steven Laureys Coma Science Group and Department of Neurology, Cyclotron Research Center, University Hospital of Liège, Liège, Belgium

Evan S. Lutkenhoff Department of Psychology, Univeristy of California Los Angeles, Los Angeles, CA, USA

Marcello Massimini Department of Biomedical and Clinical Sciences Luigi Sacco, University of Milan, Milan, Italy

Martin M. Monti Department of Psychology, University of California Los Angeles, Los Angeles, CA, USA

Department of Neurosurgery, Brain Injury Research Center, Geffen School of Medicine at University of California Los Angeles, Los Angeles, CA, USA

Lorina Naci Department of Psychology, The Brain and Mind Institute, The University of Western Ontario, London, Canada

Adrian M. Owen Department of Psychology, The Brain and Mind Institute, The University of Western Ontario, London, Canada

Francesco Piccione Fondazione Ospedale San Camillo, I.R.C.C.S., Venezia, Italy

Aldo Ragazzoni Unit of Neurology & Clinical Neurophysiology, Fondazione PAS, Scandicci, Firenze, Italy

Francesco Riganello Research in Advanced Neurorehabilitation (RAN), Institute S. Anna, Crotone, Italy

Mario Rosanova Department of Biomedical and Clinical Sciences Luigi Sacco, University of Milan, Milan, Italy

Walter G. Sannita Department of Neuroscience, Ophthalmology and Genetics, University of Genova, Genova, Italy

Department of Psychiatry, State University of New York, Stony Brook, NY, USA

Simone Sarasso Department of Biomedical and Clinical Sciences Luigi Sacco, University of Milan, Milan, Italy

Nicholas D. Schiff The Jerold B. Katz Professor of Neurology and Neuroscience, Feil Family Brain and Mind Research Institute, Weill Cornell Medical College, New York, NY, 10065, USA

Caroline Schnakers Department of Neurosurgery, Geffen School of Medicine at University of California Los Angeles, Brain Injury Research Center, Los Angeles, CA, USA

G. Bryan Young Department of Clinical Neurological Sciences, University of Western Ontario, London, ON, Canada

Chapter 1
Responsiveness in DoC: A Quest for Consciousness?

Walter G. Sannita

Abstract Interest on *consciousness* is growing, but how it arises in the brain remains fundamentally unsolved. The variety of brain mechanisms, the impairment of which is observed in DoC, has implication in neuroscience, in suggesting that *consciousness* results of a complex functional arrangement interacting with but distinct from other higher brain processes. Neuroimaging has challenged the current criteria for *consciousness* and *responsiveness* by documenting residual high-level aspects of brain activity in DoC. The clinical standards to classify patients as conscious or unconscious are questioned, and neurologists are confronted with unresolved issues about diagnosis, pain, and predictability of recovery. The epistemological issue remains the definition of *consciousness* beyond subjective feeling, verbal report, probabilistic inference, and pragmatic principles. Classification rests on *responsiveness* in the absence of a theory about *consciousness* inclusive of its quantitative characterization. Proper definitions for and an up-to-date scrutiny of the available descriptors are needed to think scientifically about *consciousness*.

Keywords Consciousness • Definition • Evolution • Descriptors of consciousness • Responsiveness • Brain functions • Clinical applications • Disorders of consciousness

1.1 Consciousness and Evolution

The eye structure (lenses plus a molecular detector of photons equal or comparable to rhodopsin) is an ancient contrivance remarkably invariant over time and across animal species. Double-structured lenses optically corrected in full accord with

W.G. Sannita, MD
Department of Neuroscience, Ophthalmology and Genetics, University of Genova, Genova, Italy

Department of Psychiatry, State University of New York, Stony Brook, NY, USA
e-mail: wgs@dism.unige.it

M.M. Monti, W.G. Sannita (eds.), *Brain Function and Responsiveness in Disorders of Consciousness*, DOI 10.1007/978-3-319-21425-2_1

constructions by Des Carts and Huygens and the laws of optics appeared with Trilobites in early Cambrian [1]. The living fossil *Limulus polyphemus* began featuring retinal mechanisms of lateral inhibition and recurrent interaction (equivalent to those allowing higher mammals to identify objects) in early Triassic, long before the development of brain structures with matching functional complexity [2][1]. The sophistication of these early components the visual system exemplifies how nature can be conservative about efficient implements even when not yet usable to their full potential.

Individual *consciousness* seems instead to have evolved late in the progression from low- to higher-order brain organization. Children spontaneously learn self-recognition in front of a mirror at about 18 mo. of age [3]; among mammals, only young chimpanzees (4.5–8 years) can be trained to this task, and the evidence with dolphins remains unclear [4]. The mirror test marks the obvious in the child, i.e., the emergence of higher brain functions and cognitive processes we associate with *consciousness*; its reliability in animal investigation is by contrast questioned. Methodological issues aside, the declining of self-recognition with adulthood in the chimpanzee suggests learning processes rather than self-awareness [3, 5]. Nonhuman primates and cetacean brains seem to have evolved to levels of complexity not too dissimilar from ours. However, the (individual and collective) adaptation to environmental requirements and the cognitive, social, and intellectual developments of humankind stand as unique, while in no other species there is evidence of conscious continuity with the past and planning for the future comparable to humans. Evidence of animal *consciousness* is intrinsically difficult to detect, and the hypothesis is often dismissed as a false problem no matter how ancient the neurophysiological processes thought to support *consciousness* in humans. Markers of evolution, nosographic criteria, and underlying physiological processes need to be unambiguously identified in approaching questions such as these, in order to qualitatively and quantitatively define *consciousness* (and its possible different states or levels) in a general taxonomy independent of self-experiencing, verbal reporting, and cultural biases [6–23].

[1] The early development of the visual system, the remarkable similarities over time and across animal species, and the loss of intermediate eye structures that appeared before the Cambrian were unknown at Charles Darwin's time. This ignorance allowed him to fully develop his theories but did not spare him troublesome concerns: "*...To suppose that the eye with all its inimitable contrivances for adjusting the focus to different distances, for admitting different amounts of light, and for the correction of spherical and chromatic aberration, could have been formed by natural selection, seems, I freely confess, absurd in the highest degree. [...] Reason tells me, that if numerous gradations from a simple and imperfect eye to one complex and perfect can be shown to exist, each grade being useful to its possessor, as is certain the case; if further, the eye ever varies and the variations be inherited, as is likewise certainly the case; and if such variations should be useful to any animal under changing conditions of life, then the difficulty of believing that a perfect and complex eye could be formed by natural selection, should not be considered as subversive of the theory...*" (Ch. Darwin, On the Origin of Species, chapter VI: Difficulties of the theory, London, John Murray, 1858).

1.2 Consciousness and Brain Functions

Research on *consciousness* focuses mostly on cortical/brain activation, network complexity, long-range connectivity, neuronal synchronization in selected frequency ranges, uni/multimodal perception, motor activation, etc. *Consciousness* is thought to result of large-scale information processing and neurobiological mechanisms [24–28], as opposed to hypotheses assimilating *consciousness* and conscious perception [29] that are based on evidence of regionally mediated synchronization between large neuronal populations of distinct areas during conscious perception [24, 30]. Sensory inflow and sensorimotor integration, brainstem activating ascending systems, thalamocortical interaction, primitive motor systems, metabolic supply, and neuromodulation balance are crucial [31]. Synchronization in the gamma-band frequency range is thought to be a mechanism for bottom-up activation of neuronal assemblies [32–37] and is reportedly maintained in disorders of consciousness (DoC) when top-down synchronization appears lost [38]. However, the contributions of these processes in sustaining *consciousness* are defined indirectly by the effects of damage. The extent to which surgical anesthesia actually affects *consciousness* remains poorly understood, with different pharmacological mechanisms of drug action apparently inducing the same main effect [39–43]. Sleep mechanisms evolved early and are ancient: the patterns of brain activation and network functional organization which sustain wakefulness, non-REM sleep, and REM sleep are incompatible to each other [44, 45], but we remain ourselves and self-conscious also in the most unrealistic dream and even happen to be aware of our dreaming [46]. The variety of brain mechanisms thought to contribute in sustaining *consciousness* (and the impairment of which results in DoC) has implication in neuroscience in suggesting that it results of a complex functional arrangement interacting with but distinct from all other higher brain processes (e.g., attention) [9, 28, 47–49]. In this respect, the master unsolved problem pervading neuroscience remains how the combination of billions of robust individual components (neurons) with flexible weak linkages between regulatory processes works together to create brain functions, including *consciousness* [50, 51].

1.3 Consciousness and Responsiveness in DoC

Neuroimaging has documented residual high-level aspects of brain activity across sensory modalities, language and learning dynamics, emotions, or pain also in subjects otherwise diagnosed as being in *vegetative state/unresponsive wakefulness syndrome* (VS/UWS) [52, 53]. Responses varied in complexity, from local activation of primary sensory cortices, to the involvement of associative areas, to activation of cortical-subcortical networks to either mental imagery tasks or distinction of ambiguous/nonambiguous words [16, 54–68]. Retained connectivity in segregated networks provides evidence of the severely damaged brain capability to express surviving modular functions that do not necessarily give rise to phenomenological awareness, in the absence of the integrative processes deemed necessary to *consciousness* [65, 69]. This has been understood as indicative of residual covert *cognition* or *consciousness* as opposed to alternative interpretations that markers of neural activity not necessarily are surrogates for these functions [63, 70–76]. The surviving networks observed in these studies compare as to anatomy and modes of activation to those observable in healthy subjects under comparable conditions. The similarity validates the hypothesis that stimulus- or condition-related regional brain activations do not occur at random in DoC. It also reinforces the evidence that neuronal and network mechanisms of the brain (mediating, e.g., in sensory data processing or motor action) can operate with limited or null interference from conscious processes [77–81].

Brain activation reflecting some awareness and cognition was observed by neuroimaging in only a small number of subjects [61]. The preservation of specific neural structures and available residual functional resources has been suggested to vary because of the heterogeneity of etiology and pathophysiology or extension and severity of brain damage [82]. Individual variability aside, the reports of subjects classified as being in VS/UWS according to clinical protocols who have proven able to produce voluntary "brain behavior" suggestive of partially retained *consciousness* during neuroimaging or neurophysiologic assessments [53, 61, 62, 83, 84] cannot be easily dismissed.

1.4 Responsiveness and the Clinical Scenario

Neurophysiology and neuroimaging have rapidly become alternative methodological approaches, and results proved scientifically seminal. These technologies have revolutionized our understanding of DoC and emphasized the possible relationship between brain response and *consciousness*. A general discussion is currently underway in the scientific and clinical communities concerning the meaning of regional brain activations in DoC, the circumstances under which brain responsiveness should be considered equivalent to *consciousness*, and the reliability of brain activations as a marker of *consciousness* [60, 85, 86]. The current definitions and traditional understanding of both responsiveness and *consciousness* are being challenged. The focus is on whether this line of scientific evidence may result misleading in research on the lower boundaries of *consciousness* [15]. It is also a matter of discussion whether patients who clinically fulfill the clinical criteria for VS/UWS but show regional brain responses should be considered conscious nonetheless and which is the acceptable discrepancy between clinical and neuroscientific evidence. According to the international guidelines [76, 87–91], the observation of each aspect of *consciousness* is based on the detection of recognizable behavioral signs, including sustained cycles of eye-opening. While practical in the clinical environment, retrospective audits and studies comparing diagnostic protocols have consistently reported a high rate of misdiagnosis between the VS/UWS and minimally conscious state (MCS) [67, 92–94]. Individual variability, either spontaneous or reflecting changes in the neuronal or non-neuronal factors modulating the brain functional state [95], appears relevant in DoC and should be taken in proper account when testing responsiveness [96, 97].

The very clinical standards by which patients are classified as conscious or unconscious have also been called into question. In the clinical community, the resulting dichotomy between traditional criteria based on observation (currently the golden standard) and evidence from advanced neuroimaging research is creating difficulties in the patient's diagnosis and early prognosis. Neurologists are confronted with unresolved issues that involve pain, sensory or emotional/affective responsiveness, and predictability of recovery and vex both the patients' families and the health team. The responsibility of diagnosis and early prognosis remains with the clinician and is mainly based on the observation of clinical responsiveness. Despite their known shortcomings [54, 59], the current behavioral markers of responsiveness have major consequences on the commitment in resources, logistics, dedicated staff, and costs needed for the management of these patients or even dramatic in ethical controversies inviting media coverage and discussion in the public forum [72, 98, 99]. VS/UWS and MCS appear today neither static nor homogeneous, and a tacit revision of the current descriptive categories is de facto underway [60, 100]. A close scrutiny of the pathophysiology of responsiveness and its relevance in diagnosis/prognosis and in the definition of the boundaries between reactive VS/UWS and MCS is mandatory if scientific inquiry is to resolve the practical problem of determining a patient's state of *consciousness*. In an effort to address this issue, the clinical and scientific communities have begun evaluating the potential usefulness of novel technologies to integrate and supplement standard clinical procedures and to investigate the mechanisms underlying the loss and recovery of *consciousness* [19, 43, 62, 83, 84, 90, 101–111].

1.5 Consciousness and Its Descriptors

Interest on *consciousness* is growing in neuroscience and medicine as well as in neurocomputing, artificial intelligence, and robotics with the rapid progress in the investigation of higher brain functions, advance in artificial intelligence, and diffuse perception of the inadequacy of traditional mind/body separations [8, 10, 112–114]. As a result, the concept of *consciousness* may vary depending on context, scientific approach, and background (from neurophysiological state of activation to self-awareness, to momentary interaction with the environment, etc.), and the definitions of *consciousness* and related terms remain to a significant extent inadequately characterized and ambiguous. The binary classification of *consciousness* vs. unconsciousness is being questioned on the ground of the neurophysiological correlates of preconscious processes [115–117], whereas the evidence for non-unitary models remains not definite [15]. The main epistemological issue about *consciousness* in humans and possibly in other species remains its definition beyond subjective feeling, verbal report, probabilistic inference (since by definition anyone's own experience is inaccessible to others), and pragmatic principle of "revealed consciousness" [59, 60].

How *consciousness* arises in the brain remains fundamentally unsolved. It has been noted in his respect how research relies to a relevant extent on the linguistic neutrality of neurophysiological and pathophysiological "correlates" when the experimental paradigms and explanatory canons are not neutral about the mechanical relations with the brain and are supposed to investigate causes [118]. Neuroscience has advanced to the point that *consciousness* seems treatable as a scientific problem like others, disregarding objections that it may be epiphenomenal, not evolutionary in function, unaccountable by brain processes, intrinsically unsuitable to objective investigated, etc. [112]. Yet, a classification of DoC remains bound to the idea of responsiveness until a theory of consciousness inclusive of its quantitative characterization emerges from current scientific proposals. To this end, proper definitions for *consciousness* and responsiveness and an up-to-date scrutiny of the descriptors are needed in order to think scientifically about *consciousness* and start experimental studies.

References

1. Clarkson EN, Levi-Setti R (1975) Trilobite eyes and the optics of Des Cartes and Huygens. Nature 254(5502):663–667
2. Wade NJ (2009) Hartline's Limulus. Cortex 45(6):695–696. doi:10.1016/j.cortex.2008.02.010. Epub 2008 Nov 19
3. Povinelli DJ, Rulf AB, Landau KR, Bierschwale DT (1993) Self-recognition in chimpanzees (Pan troglodytes): distribution, ontogeny, and patterns of emergence. J Comp Psychol 107(4): 347–372
4. Harley HE (2013) Consciousness in dolphins? A review of recent evidence. J Comp Physiol A Neuroethol Sens Neural Behav Physiol 199(6):565–582
5. Rochat P, Zahavi D (2011) The uncanny mirror: a re-framing of mirror self-experience. Conscious Cogn 20(2):204–213. doi:10.1016/j.concog.2010.06.007. Epub 2010 Oct 2
6. Boly M, Seth AK, Wilke M, Ingmundson P, Baars B, Laureys S, Edelman DB, Tsuchiya N (2013) Consciousness in humans and non-human animals: recent advances and future directions. Front Psychol 4:625. Review
7. Cook ND, Carvalho GB, Damasio A (2014) From membrane excitability to metazoan psychology. Trends Neurosci 37:698–705, pii: S0166-2236(14)00128-3
8. Damasio A (1999) The feeling of what happens: body and emotion in the making of consciousness. Harcourt Press, New York
9. Dienes Z, Seth AK (2010) Measuring any conscious content versus measuring the relevant conscious content: comment on Sandberg et al. Conscious Cogn 19:1079–1080
10. Edelman GM (2003) Naturalizing consciousness: a theoretical framework. Proc Natl Acad Sci U S A 100:5520–5524
11. Edelman DB, Seth AK (2009) Animal consciousness: a synthetic approach. Trends Neurosci 32(9):476–484
12. Griffin D (2001) Animal minds: beyond cognition to consciousness. University of Chicago Press, Chicago
13. Hofstadter D. Reflections on what is it like to be a bat? In: Hofstadter D, Dennett D (eds) The Mind's I. Basic Books. pp 403–414
14. Kotchoubey B (2014) Objectivity of human consciousness is a product of tool usage. Front Psychol 5:1152. doi:10.3389/fpsyg.2014.01152. eCollection 2014. PubMed PMID: 25346714, PubMed Central PMCID: PMC4191348

15. Kotchoubey B, Vogel D, Lang S, Müller F (2014) What kind of consciousness is minimal? Brain Inj 28(9):1156–1163
16. Laureys S, Piret S, Ledoux D (2005) Quantifying consciousness. Lancet Neurol 4(12): 789–790
17. Mashour GA, Alkire MT (2013) Evolution of consciousness: phylogeny, ontogeny, and emergence from general anesthesia. Proc Natl Acad Sci U S A 110(Suppl 2):10357–10364
18. Nagel T (1991) Mortal questions, vol 12, What is it like to be a bat? Cambridge University Press, Cambridge, UK
19. Overgaard M, Timmermans B, Sandberg K, Cleeremans A (2010) Optimizing subjective measures of consciousness. Conscious Cogn 19(2):682–684; discussion
20. Schneider S, Velmans M (2008) Introduction. In: Velmans M, Schneider S. (eds) The Blackwell companion to consciousness. Wiley, Malden/Oxford
21. Searle JR (1998) How to study consciousness scientifically. Philos Trans R Soc Lond B Biol Sci 353(1377):1935–1942
22. Seth AK, Dienes Z, Cleeremans A, Overgaard M, Pessoa L (2008) Measuring consciousness: relating behavioural and neurophysiological approaches. Trends Cogn Sci 12(8):314–321. Review
23. Velmans M (2009) How to define consciousness—and how not to define consciousness. J Conscious Stud 16:139–156
24. Koch C, Laurent G (1999) Complexity and the nervous system. Science 284:96–98
25. Moreno D, Schiff ND, Giacino J, Kalmar K, Hirsch J (2010) A network approach to assessing cognition in disorders of consciousness. Neurology 75:1871–1978
26. Tononi G (2005) Consciousness, information integration, and the brain. Prog Brain Res 150:109–126
27. Tononi G, Edelman GM (1998) Consciousness and complexity. Science 282:1846–1851
28. Tononi G, Koch C (2008) The neuronal correlates of consciousness: an update. Ann N Y Acad Sci 1124:239–362
29. Zeki S (2003) The disunity of consciousness. Trends Cogn Sci 7:214–218
30. Srinivasan R, Russell DP, Edelman GM, Tononi G (1999) Increased synchronization in neuromagnetic responses during conscious perception. J Neurosci 19:5435–5438
31. Dolce G, Sazbon L (eds) (2002). The posttraumatic vegetative state. Stuttgart, Thiene
32. Bressler SL (1990) The gamma wave: a cortical information carrier? Trends Neurosci 13: 161–162
33. Engel AK, König P, Kreiter AK, Schillen TB, Singer W (1992) Temporal coding in the visual cortex: new vistas on integration in the nervous system. Trends Neurosci 15:218–226
34. Jefferys JG, Traub RD, Whittington MA (1996) Neuronal networks for induced '40 Hz' rhythms. Trends Neurosci 19:202–208
35. Sannita WG (2000) Stimulus-specific oscillatory responses of the brain: a time/frequency-related coding process. Clin Neurophysiol 111:565–583
36. Singer W (1993) Synchronization of cortical activity and its putative role in information processing and learning. Ann Rev Physiol 55:349–374
37. Traub RD, Jefferys JGR, Whittington MA (1999) Fast oscillations in cortical circuits. Computational neuroscience. MIT Press, Cambridge, MA
38. Boly M, Garrido MI, Gosseries O, Bruno MA, Boveroux P, Schnakers C, Massimini M, Litvak V, Laureys S, Friston K (2011) Preserved feedforward but impaired top-down processes in the vegetative state. Science 332(6031):858–862
39. Alkire MT, Hudetz AG, Tononi G (2008) Consciousness and anesthesia. Science 322(5903): 876–880
40. Boly M, Sanders RD, Mashour GA, Laureys S (2013) Consciousness and responsiveness: lessons from anaesthesia and the vegetative state. Curr Opin Anaesthesiol 26(4): 444–449
41. MacDonald AA, Naci L, MacDonald PA, Owen AM (2015) Anesthesia and neuroimaging: investigating the neural correlates of unconsciousness. Trends Cogn Sci 19(2):100–107
42. Mashour GA, Alkire MT (2013) Consciousness, anesthesia, and the thalamocortical system. Anesthesiology 118(1):13–15

43. Monti MM, Lutkenhoff ES, Rubinov M, Boveroux P, Vanhaudenhuyse A, Gosseries O, Bruno MA, Noirhomme Q, Boly M, Laureys S (2013) Dynamic change of global and local information processing in propofol-induced loss and recovery of consciousness. PLoS Comput Biol 9(10):e1003271
44. Huber R, Ghilardi MF, Massimini M, Tononi G (2004) Local sleep and learning. Nature 430:78–81
45. Maquet P (2000) Functional neuroimaging of normal human sleep by positron emission tomography. J Sleep Res 9:207–231
46. Kahn D, Gover T (2010) Consciousness in dreams. Int Rev Neurobiol 92:181–195
47. Koch C, Tsuchiya N (2008) Attention and consciousness: two distinct brain processes. Trends Cogn Sci 12:44
48. van Boxtel JJ, Tsuchiya N, Koch C (2010) Opposing effects of attention and consciousness on afterimages. Proc Natl Acad Sci U S A 107(19):8883–8888
49. Zeman A (2001) Consciousness. Brain 124:1263–1289. Review
50. Arecchi FT, Meucci R, Sannita WG, Farini A (guest eds) (2009) Neuroscience today: neuronal functional diversity and collective behaviors. Cogn Proc Suppl 1
51. Sannita WG (2008) Neuronal functional diversity and collective behaviors. J Biol Phys 34:267–278
52. Jennett B (2002) The vegetative state. University Press, Cambridge, UK
53. Laureys S, Celesia GG, Cohadon F, Lavrijsen J, León-Carrión J, Sannita WG, Sazbon L, Schmutzhard E, von Wild KR, Zeman A, Dolce G (2010) European Task Force on Disorders of Consciousness. Unresponsive wakefulness syndrome: a new name for the vegetative state or apallic syndrome. BMC Med 8:68
54. Coleman MR, Rodd JM, Davis MH, Johnsrude IS, Menon DK, Pickard JD, Owen AM (2007) Do vegetative patients retain aspects of language comprehension? Evidence from fMRI. Brain 130(Pt 10):2494–2507
55. Coleman MR, Davis MH, Rodd JM, Robson T, Ali A, Owen AM, Pickard JD (2009) Towards the routine use of brain imaging to aid the clinical diagnosis of disorders of consciousness. Brain 132(Pt 9):2541–2552
56. Kotchoubey B, Lang S, Mezger G, Schmalohr D, Schneck M, Semmler A, Bostanov V, Birbaumer N (2005) Information processing in severe disorders of consciousness: vegetative state and minimally conscious state. Clin Neurophysiol 116:2441–2453
57. Laureys S, Owen AM, Schiff ND (2004) Brain function in coma, vegetative state, and related disorders. Lancet Neurol 3(9):537–546. Review
58. Laureys S, Perrin F, Schnakers C, Boly M, Majerus S (2005) Residual cognitive function in comatose, vegetative and minimally conscious states. Curr Opin Neurol 18:726–733. Review
59. Monti MM, Coleman MR, Owen AM (2009) Neuroimaging and the vegetative state: resolving the behavioral assessment dilemma? Ann N Y Acad Sci 1157:81–89
60. Monti MM, Laureys S, Owen AM (2010) The vegetative state. BMJ 341:c3765. doi:10.1136/bmj.c3765. Review
61. Monti MM, Vanhaudenhuyse A, Coleman MR, Boly M, Pickard JD, Tshibanda L, Owen AM, Laureys S (2010) Willful modulation of brain activity in disorders of consciousness. N Engl J Med 362(7):579–589
62. Owen AM, Coleman MR, Boly M, Davis MH, Laureys S, Pickard JD (2006) Detecting awareness in the vegetative state. Science 313(5792):1402
63. Owen AM, Coleman MR (2008) Using neuroimaging to detect awareness in disorders of consciousness. Funct Neurol 23(4):189–194. Review
64. Riganello F, Sannita WG (2009) Residual brain processing in vegetative state. J Psychophysiol 23:18–26. Review
65. Schiff ND, Ribary U, Moreno DR, Beattie B, Kronberg E, Blasberg R, Giacino J, McCagg C, Fins JJ, Llinás R, Plum F (2002) Residual cerebral activity and behavioral fragments can remain in the persistently vegetative brain. Brain 125(6):1210–1234
66. Schiff ND (2005) Modeling the minimally conscious state: measurements of brain function and therapeutic possibilities. Prog Brain Res 150:473–493. Review

67. Schnakers C, Vanhaudenhuyse A, Giacino J, Ventura M, Boly M, Majerus S, Moonen G, Laureys S (2009) Diagnostic accuracy of the vegetative and minimally conscious state: clinical consensus versus standardized neurobehavioral assessment. BMC Neurol 9:35
68. Schnakers C, Chatelle C, Demertzi A, Majerus S, Laureys S (2012) What about pain in disorders of consciousness? AAPS J 14(3):437–444
69. Plum F, Schiff N, Ribary U, Llinás R (1998) Coordinated expression in chronically unconscious persons. Philos Trans R Soc Lond B Biol Sci 353(1377):1929–1933
70. Bardin J, Fins JJ, Katz DI, Hersh J, Heier LA, Tabelow K, Dyke JP, Ballon DJ, Schiff ND, Voss H (2011) Dissociations between behavioural and functional magnetic resonance imaging-based evaluations of cognitive function after brain injury. Brain 134:769–782
71. Celesia GG (2013) Conscious awareness in patients in vegetative states: myth or reality? Curr Neurol Neurosci Rep 13(11):395. Review
72. Celesia GG, Sannita WG (2013) Can patients in vegetative state experience pain and have conscious awareness? Neurology 80(4):328–329. Editorial
73. King JR, Sitt JD, Faugeras F, Rohaut B, El Karoui I, Cohen L, Naccache L, Dehaene S (2013) Information sharing in the brain indexes consciousness in noncommunicative patients. Curr Biol 23(19):1914–1919
74. Nachev P, Hacker PM (2010) Covert cognition in the persistent vegetative state. Prog Neurobiol 91(1):68–76
75. Schwartz R, Schwartz M (2008) The risks of reducing consciousness to neuroimaging. Am J Bioeth 8(9):25–26
76. Turner-Stokes L, Kitzinger J, Gill-Thwaites H, Playford ED, Wade D, Allanson J, Pickard J (2012) fMRI for vegetative and minimally conscious states. A more balanced perspective. BMJ 345:e8045
77. Destrebecqz A, Peigneux P (2006) Methods for studying unconscious learning. In: Laureys S (ed) The boundaries of consciousness: neurobiology and neuropathology. Elsevier, Amsterdam
78. Radoeva PD, Prasad S, Brainard DH, Aguirre GK (2008) Neural activity within area V1 reflects unconscious visual performance in a case of blindsight. J Cogn Neurosci 20:1927–1939
79. Sannita WG (2014) Human brain physiology investigated in the disorder of consciousness. Front Neurol 5:211
80. Sergent C, Baillet S, Dehaene S (2005) Timing of the brain events underlying access to consciousness during the attentional blink. Nat Neurosci 8:1391–1400
81. Sklar AY, Levy N, Goldstein A, Mandel R, Maril A, Hassin RR (2012) Reading and doing arithmetic nonconsciously. Proc Natl Acad Sci U S A 109:19614–19619
82. Gibson RM, Fernández-Espejo D, Gonzalez-Lara LE, Kwan BY, Lee DH, Owen AM, Cruse D (2014) Multiple tasks and neuroimaging modalities increase the likelihood of detecting covert awareness in patients with disorders of consciousness. Front Hum Neurosci 8:950
83. Lutkenhoff ES, Chiang J, Tshibanda L, Kamau E, Kirsch M, Pickard JD, Laureys S, Owen AM, Monti MM (2015) Thalamic and extrathalamic mechanisms of (un)consciousness after severe brain injury. Ann Neurol 20
84. Monti MM, Rosenberg M, Finoia P, Kamau E, Pickard JD, Owen AM (2015) Thalamo-frontal connectivity mediates top-down cognitive functions in disorders of consciousness. Neurology 84(2):167–173
85. Bernat JL (2002) Questions remaining about the minimally conscious state. Neurology 58(3):337–338
86. Stender J, Gosseries O, Bruno MA, Charland-Verville V, Vanhaudenhuyse A, Demertzi A, Chatelle C, Thonnard M, Thibaut A, Heine L, Soddu A, Boly M, Schnakers C, Gjedde A, Laureys S (2014) Diagnostic precision of PET imaging and functional MRI in disorders of consciousness: a clinical validation study. Lancet 384(9942):514–522
87. Fins JJ, Master MG, Gerber LM, Giacino JT (2007) The minimally conscious state: a diagnosis in search of an epidemiology. Arch Neurol 64(10):1400–1405. Review
88. Giacino JT (2005) The minimally conscious state: defining the borders of consciousness. Prog Brain Res 150:381–395. Review

89. Giacino JT, Malone R (2008) The vegetative and minimally conscious states. Handb Clin Neurol 90:99–111. doi:10.1016/S0072-9752(07)01706-X. Review
90. Giacino JT, Fins JJ, Laureys S, Schiff ND (2014) Disorders of consciousness after acquired brain injury: the state of the science. Nat Rev Neurol 10(2):99–114
91. Multi-Society Task Force on PVS (1994) Statement on medical aspects of the persistent vegetative state. N Engl J Med 330:1499–1508
92. Andrews K, Murphy L, Munday R, Littlewood C (1996) Misdiagnosis of the vegetative state: retrospective study in a rehabilitation unit. BMJ 313:13–16
93. Bosco A, Lancioni GE, Belardinelli MO, Singh NN, O'Reilly MF, Sigafoos J (2010) Vegetative state: efforts to curb misdiagnosis. Cogn Process 11:87–90
94. Goldfine AM, Bardin JC, Noirhomme Q, Fins JJ, Schiff ND, Victor JD (2013) Reanalysis of bedside detection of awareness in the vegetative state: a cohort study. Lancet 381(9863):289–291
95. Sannita WG (2006) Individual variability, end-point effects and possible biases in electrophysiological research. Clin Neurophysiol 117:2569–2583
96. Candelieri A, Cortese MD, Dolce G, Riganello F, Sannita WG (2011) Visual pursuit: within-day variability in the severe disorder of consciousness. J Neurotrauma 28:1149–1154
97. Cortese MD, Riganello F, Arcuri F, Pugliese ME, Lucca LF, Dolce G, Sannita WG (2015) Coma recovery scale-r: variability in disorders of consciousness. BMC Neurology 14:186
98. Johnson LS (2011) The right to die in the minimally conscious state. J Med Ethics 37(3):175–178
99. Racine E, Amaram R, Seidle M, Karczewska M, Illes J (2008) Media coverage of the persistent vegetative state and end-of-life decision-making. Neurology 71:1027–1032
100. Dolce G, Sannita WG (2010) The vegetative state: a syndrome seeking revision? Brain Inj 24:1628–1629
101. Adams JH, Graham DI, Jennett B (2000) The neuropathology of the vegetative state after an acute brain insult. Brain 123(Pt 7):1327–1338
102. Chennu S, Finoia P, Kamau E, Allanson J, Williams GB, Monti MM, Noreika V, Arnatkeviciute A, Canales-Johnson A, Olivares F, Cabezas-Soto D, Menon DK, Pickard JD, Owen AM, Bekinschtein TA (2014) Spectral signatures of reorganised brain networks in disorders of consciousness. PLoS Comput Biol 10(10):e1003887
103. Crone JS, Soddu A, Höller Y, Vanhaudenhuyse A, Schurz M, Bergmann J, Schmid E, Trinka E, Laureys S, Kronbichler M (2013) Altered network properties of the fronto-parietal network and the thalamus in impaired consciousness. Neuroimage Clin 4:240–248
104. Crone JS, Schurz M, Höller Y, Bergmann J, Monti M, Schmid E, Trinka E, Kronbichler M (2015) Impaired consciousness is linked to changes in effective connectivity of the posterior cingulate cortex within the default mode network. Neuroimage 110:101–109. doi:10.1016/j.neuroimage.2015.01.037. Epub 2015 Jan 22
105. Demertzi A, Gómez F, Crone JS, Vanhaudenhuyse A, Tshibanda L, Noirhomme Q, Thonnard M, Charland-Verville V, Kirsch M, Laureys S, Soddu A (2014) Multiple fMRI system-level baseline connectivity is disrupted in patients with consciousness alterations. Cortex 52:35–46. doi:10.1016/j.cortex.2013.11.005. Epub 2013 Nov 20. PubMed
106. Fernández-Espejo D, Owen AM (2013) Detecting awareness after severe brain injury. Nat Rev Neurosci 14(11):801–809
107. Gómez F, Phillips C, Soddu A, Boly M, Boveroux P, Vanhaudenhuyse A, Bruno MA, Gosseries O, Bonhomme V, Laureys S, Noirhomme Q (2013) Changes in effective connectivity by propofol sedation. PLoS One 8(8):e71370. doi:10.1371/journal.pone.0071370. eCollection 2013. PubMed PMID: 23977030, PubMed Central PMCID: PMC3747149
108. Lutkenhoff ES, Rosenberg M, Chiang J, Zhang K, Pickard JD, Owen AM, Monti MM (2014) Optimized brain extraction for pathological brains (optiBET). PLoS One 9(12):e115551
109. Naci L, Monti MM, Cruse D, Kübler A, Sorger B, Goebel R, Kotchoubey B, Owen AM (2012) Brain-computer interfaces for communication with nonresponsive patients. Ann Neurol 72(3):312–323
110. Naci L, Owen AM (2013) Making every word count for nonresponsive patients. JAMA Neurol 70(10):1235–1241

111. Naci L, Cusack R, Jia VZ, Owen AM (2013) The brain's silent messenger: using selective attention to decode human thought for brain-based communication. J Neurosci 33(22):9385–9393
112. Crick F, Koch C (1998) Consciousness and neuroscience. Cereb Cortex 8:97–107. Review
113. Dennett D (1991) Consciousness explained. Little & Company, Boston
114. Koch C (2004) The Quest for consciousness. Roberts and Co., Englewood
115. Dehaene S, Changeux JP, Naccache L, Sackur J, Sergent C (2006) Conscious, preconscious, and subliminal processing: a testable taxonomy. Trends Cogn Sci 10:204–211
116. Marois R, Yi DJ, Chun MM (2004) The neural fate of consciously perceived and missed events in the attentional blink. Neuron 41:465–472
117. Sitt JD, King JR, Naccache L, Dehaene S (2013) Ripples of consciousness. Trends Cogn Sci 17(11):552–554
118. Neisser J (2012) Neural correlates of consciousness reconsidered. Conscious Cogn 21(2):681–690

Chapter 2
Advances in the Scientific Investigation of Consciousness

Daniel Bor

Abstract Questions of consciousness lie at the heart of many ethical debates, including the withdrawal of life support from DoC patients. Therefore understanding the nature of consciousness is one of the most vital tasks remaining in biological science. Thanks in part to new neuroimaging techniques, considerable progress has been made in understanding consciousness science over the last two decades. At the behavioural level, attention is a necessary component of consciousness, and most complex mental processes require consciousness to function. However, consciousness is severely capacity limited to 3–4 items. Given these features, consciousness might serve an evolutionary function to provide innovative solutions to survival-critical problems that instincts or habits could not solve. Functional imaging and focal lesion patient studies have linked conscious contents to the lateral prefrontal cortex and posterior parietal cortex. The manipulation of conscious levels, such as between wakefulness and sedation, shows a reduction in the same brain regions. Although these regions are also associated in DoC patients, thalamus damage is a more common marker of vegetative state. Current theories posit the importance of network architecture and dynamics, such as recurrent processing, for supporting consciousness and closely link integrated information with consciousness. The most promising current imaging assessments for clinical adoption, reflecting this theoretical perspective, use novel analysis methods to gauge the informational complexity of the neural signal, as a marker of residual conscious level.

Keywords Consciousness • Attention • Working memory • Prefrontal cortex • Parietal cortex • Disorders of consciousness • Vegetative state

---

D. Bor
Sackler Centre for Consciousness Science, University of Sussex, Brighton, UK

Department of Informatics, University of Sussex, Brighton BN1 9QJ, UK
e-mail: D.Bor@sussex.ac.uk

M.M. Monti, W.G. Sannita (eds.), *Brain Function and Responsiveness in Disorders of Consciousness*, DOI 10.1007/978-3-319-21425-2_2

2.1 Introduction

Consciousness is required for all meaningful experiences. Consequently, many ethical questions, including the withdrawal of life support in disorders of patients with DoC, are underpinned by that person's capacity for consciousness. Unlocking the nature of consciousness, therefore, is one of the most profound issues remaining in biology [1, 2].

Defining consciousness precisely has proven difficult [3]. However, this has not limited scientific progress, since broad features of consciousness can be safely assumed and explored, such as the presence of phenomenal content for an agent. From this skeletal description, one useful distinction that arises is between the degree of presence of phenomenal content and the range of phenomenal states available. In other words, *conscious level* (with coma an example of the lowest level, involving complete unconsciousness, and alert wakefulness reflecting the highest levels) is compared with *conscious content* (known in philosophical circles as *qualia* or the actual subjective components of any particular experience). At least some positive conscious level is necessary for conscious contents to occur.

Over the past two decades, increasing scientific attention has been focused on the study of consciousness, such that it is now a very active and sophisticated research subject. Consciousness science currently employs a wide range of techniques, including behavioural studies, computational models, neuropsychology, transcranial magnetic stimulation (TMS), magneto- and electroencephalography (M/EEG), structural and functional magnetic resonance imaging (fMRI), intracranial recordings, and electrophysiology [4]. In addition to research examining changes in conscious content for awake subjects, a range of conscious levels have been explored in normal participants, including alert wakefulness, sleep, and sedation. Furthermore, these techniques are commonly used on various patient groups with relevance to consciousness science, such as those with locked in syndrome, minimally conscious state (MCS) patients, vegetative state (VS) patients, and those in a coma [5].

In this chapter I will describe the empirical progress that has been made in the science of consciousness, first from a behavioural and then from a neurophysiological perspective. I will then outline how these results have given rise to current neural theories of consciousness and finally how these models pertain to studying DoC patients.

2.2 The Psychology of Consciousness

Early psychological descriptions of consciousness closely associated it with attention and emphasised its role in serial processing as a 'limited capacity channel' [6]. Later historical models of consciousness identified conscious contents with working memory contents [7] and executive control [8].

The most prominent psychological theory of consciousness, 'global workspace theory' [9, 10], echoes these earlier perspectives. The theory proposes that there is an unconscious competitive process among local units, in order to win access to a 'global workspace' of widely available content. The global broadcasting of these contents to multiple receiving modules establishes their presence in consciousness. This theory, with elements both of working memory and attention, provides some explanation for the unified and serial nature of consciousness, but fails adequately to explain the phenomenology of awareness.

The question of the association between consciousness and attention has been a source of vigorous recent debate [11–14]. At the turn of the twenty-first century, it was widely assumed that attention and consciousness were at the very least very closely related [15, 16]. However, over the last decade, there have been claims that consciousness and attention are entirely dissociable and are, in fact, independent processes, with attention neither being necessary nor sufficient for consciousness [14].

Although such positions are entirely dependent on exact definitions of consciousness and attention, there is nevertheless good evidence that attentional effects can occur without the attended stimuli reaching consciousness [17–21]. For example, the response times to a visible target were modulated by the congruency of invisible prime locations that had just been attended to [19]. Although such studies do indeed demonstrate that attention can be effectively deployed on stimuli that are not consciously perceived, it does not necessarily follow that consciousness is dissociated from attention. It is possible, for instance, that attention towards an unconscious object increases the probability that the object will be consciously perceived. Such an interpretation would be consistent with other studies that have demonstrated that attention increases conscious detection rates [22] and enhances the contrast of a consciously perceived stimulus [23].

Evidence for the opposite situation, of consciousness without attention, is far weaker and more controversial. For instance, natural scene perception, or 'gist', where general semantic details of a visual scene are consciously available, even when presentation times are extremely brief [24], is used to argue that consciousness can occur without attention. The argument goes that if meaningful consciousness is possible with a presentation time of 150 ms, then consciousness does not require attention, since this timeframe is too short for attention to be established [14]. However, it seems entirely possible that selective attention, which requires about 300 ms to occur [25], can easily act on iconic memory, which persists for longer than the duration of stimulus presentation. In line with this, if attention is effectively withdrawn from a briefly presented visual scene, then visibility of the gist of the scene is also largely abolished [26, 27]. Indeed, for gist or any other claimed example of consciousness without top-down attention, either bottom-up attention is most likely present or, when experiments have actively removed attentional resources, consciousness for a given stimulus has also been reduced or removed [11]. Therefore, the weight of evidence suggests that while attention can occur without consciousness, the opposite is not the case.

Attention is therefore probably a necessary component of consciousness, acting as the gateway for conscious contents [1, 11, 28]. Many lines of evidence support the important role of attention for consciousness. Inattentional blindness is a well-studied paradigm where attention is drawn away from an otherwise striking and perfectly visible object (such as a gorilla walking across a scene), causing the subject to fail to consciously perceive it [29, 30]. Change blindness is a similar paradigm, where in its usual form there is an obvious change between two otherwise identical visual images, which alternate, interleaved by a brief blank screen. Again, if attention isn't directed at this clear difference, the subject fails consciously to detect the change [31]. Both these paradigms demonstrate the surprising limits of conscious contents, where only a small subset of available objects can be simultaneously perceived, presumably constrained by short-term memory capacity limits and therefore only allowing a maximum of approximately 3–4 items [32]. However, these items are usually only consciously perceived as fully formed, high-level mental objects, integrated across modalities and carrying all relevant conceptual and linguistic content.

Further links between attention and consciousness are demonstrated by the neurological condition of hemispatial neglect, which usually results from damage to the right parietal or prefrontal cortex [33, 34]. Although classically assumed to be an attentional deficit, neglect is just as easily characterised as a reduction of awareness of one side of space, most frequently on the left. Neglect is neither a condition of sensory impairment per se nor is it limited to the visual modality. For instance, touch-based neglect has been reported [35].

In addition to links with attention and working memory, consciousness has been associated generally with effortful cognitive processing [36] and has been shown to be required for most complex or novel forms of thought, including understanding cause and effect, any nontrivial mathematical tasks, most logical operations, sequential information, as well as processing and acting on social knowledge [37].

In line with consciousness being intimately associated with complex information processing, it also appears to be heavily modulated by prior expectations [38, 39]. This view of consciousness as an attentionally gated, highly limited mental workspace suitable for fully processed mental items, where novel and complex processes are carried out, suggests a candidate evolutionary role for consciousness: it may provide a mental space where disparate forms of information can be compared and combined, in order to discover innovative solutions to novel or complex problems [1, 11]. In this way, more advanced strategies can be formulated to reach otherwise intractable biological goals and avoid otherwise probable dangers.

2.3 The Neurophysiology of Consciousness

Pioneering early studies to examine the neural correlates of conscious vision used single-unit recording in monkeys as they performed a binocular rivalry task, where one image was presented to the left eye and a different image was presented to the right [40, 41]. Although these images didn't change, the percept semi-regularly switched, as it does in humans. Logothetis and colleagues found that whereas only about 20 % of the primary and secondary visual cortex neurons tracked changing perceptions, almost all inferotemporal cortex neurons were excited by the percept itself. This leads to the suggestion that late ventral stream object recognition neurons were somehow responsible for consciousness. However, since this technique only sampled the visual stream itself, it was unclear at the time whether other brain regions were equally as involved in conscious processes. Indeed, in an extension of this work, again using monkey single-unit recording and a version of binocular rivalry, it was found that an equally high proportion of lateral prefrontal cortex neurons responded to the percept itself, indicating that this region is also heavily involved in consciousness [42]. This result highlights the predominant finding in consciousness research, which is the involvement in conscious processing not just of sensory specific regions but also of association cortex. From many different paradigms, the most robust observation in the literature has been the link between consciousness and the lateral prefrontal cortex, along with the posterior parietal cortex, often referred to together as the prefrontal parietal network (PPN). Human fMRI studies that have used bistable percepts, such as binocular rivalry or ambiguous figures (e.g. the Necker cube), have consistently observed PPN activations when percepts have switched from one stimulus viewpoint to the other [28, 43, 44].

Another way of examining the neural correlates of conscious content is to manipulate conscious detection of the same type of stimuli, for instance, by masking. In one study, using visual words, undetected stimuli were largely associated with the visual word form area. However, when the stimuli were visible, visual word form area activity was significantly raised, and new activity was observed in the PPN [45]. Analogous results have been observed in the auditory [46] and tactile domains [47], where large-scale PPN activity, usually bilaterally, is only associated with awareness of the stimulus. One issue with such studies is that they could be confounding subjective visibility with objective performance. To address this issue, Lau and Passingham used a metacontrast masking study, where objective performance was matched, whether the stimulus was visible or not [48]. Using this paradigm, they observed that only the dorsolateral prefrontal cortex (DLPFC) activity was linked with visible stimuli. Although these studies fail to provide a causal role for the PPN in conscious processing, there is a wealth of data from other sources to indicate that the PPN is indeed necessary for consciousness. For instance, unilateral prefrontal cortex lesion patients were less likely than controls to consciously detect briefly presented numbers, when controlling for objective performance [49]. In addition, a rare bilateral prefrontal lesion patient was reported to be awake, but almost completely unresponsive to stimuli [50].

Deficits in consciousness have also been reported following posterior parietal cortex lesions. For instance, bilateral posterior parietal patients showed reduced subjective experience in a recollection task, when compared with controls, despite matched objective performance [51]. The PPN has also been shown to be necessary for consciousness in normal subjects, via transcranial magnetic stimulation (TMS). For instance, TMS either applied to the DLPFC or the posterior parietal cortex impairs conscious change detection [52, 53]. In addition, Rounis and colleagues reported impaired visibility ratings, when controlling for objective performance, after TMS was applied to the DLPFC [54]. In a similar study, which also controlled for objective performance, Fleming and colleagues found that subjective ratings of awareness of the stimulus correlated with the anterior lateral prefrontal grey matter density [55]. In a follow-up study, the authors demonstrated a correlation between activity in this region and reported confidence [56].

Structural changes in the parietal cortex have also been linked with consciousness. For instance, the individual differences in perceptual alternation rates for bistable moving images correlated with posterior parietal grey matter density, cortical thickness, and white matter integrity [57]. The links between consciousness level and the PPN are equally robust. Bilateral damage to parietal and prefrontal white matter is associated with coma or VS [58]. Furthermore, disrupted backward connectivity between the prefrontal and temporal cortices was found in VS, as compared with controls [59]. Similarly, intact functional connectivity between the prefrontal cortex and the thalamus is predictive of recovery from vegetative state [60]. If the PPN is critically involved in conscious processes, one would also assume that PPN activity would reduce as conscious levels drop. And this has indeed been observed in a range of studies, for instance, comparing normal awake subjects with those in a slow-wave sleep [61] or sedated following the administration of general anaesthesia [62, 63]. In addition, VS patients have specific reductions in PPN activity compared with controls [64].

The DoC field, in addition to the PPN, also highlights the critical role of the thalamus. For instance, thalamic abnormalities are the most common neurophysiological dysfunction in vegetative state patients [65]. Reinforcing this point, Fernandez-Espejo and colleagues found that lack of white matter integrity in the thalamus was the main diagnostic marker for vegetative state, compared with minimally conscious state [66]. In addition, Schiff and colleagues showed that thalamic deep brain stimulation in a minimally conscious state patient significantly improved conscious levels and general functional outcomes [67].

The emerging picture from all these neurophysiological studies is that content-specific regions provide the conscious details (e.g. the inferotemporal cortex for visual objects) but that for all manner of conscious contents, the PPN in concert with the thalamus are required for consciousness to occur. Why should these regions be so closely associated with consciousness? The PPN is generally implicated in novel and complex processing [11, 68] and is particularly robustly activated when performing high-level strategic processing, even compared with more difficult versions of the same task [69–71]. The thalamus is thought to act as a multisensory information relay hub and therefore could be seen to facilitate the formation of multimodal mental objects, primarily within the PPN. These functional roles are quite consistent with the view outlined in the psychological section above, where a potential evolutionary purpose for consciousness might be to discover innovative solutions to otherwise intractable problems.

2.4 Neural Theories and Models of Consciousness

Given the relatively consistent and mature empirical picture, the main current theories of consciousness are broadly in agreement over certain key features of how consciousness could be physically instantiated. One assumption common to theories is that feedforward activity alone is not sufficient for consciousness to occur; instead, recurrent processing is required, preferably in a global way that encompasses the PPN [72]. Current theories also emphasise that network architecture is key to supporting consciousness, with central, highly interconnected regions better able to generate consciousness and those more peripheral, isolated areas only providing a support role at best, possibly by supplying specific conscious contents [28, 73]. This view is consistent with various anatomical features relevant to consciousness, such as that the PPN is a substantial portion of a particularly densely interconnected 'inner core' of brain networks, as measured by monkey white matter tracing [74]. In addition, the cerebellum, despite having considerably more neurons than the entire cortex, is not necessary for consciousness, possibly because it is largely composed of small independent, isolated modules [73].

The two most popular current physical models of consciousness, global neuronal workspace theory (GNW) and integrated information theory (IIT), also both emphasise two features of a network that are reflective of consciousness: global integration of information in a network and the ability of that network to represent a large repertoire of different functional states [75, 28, 76]. GNW diverges from IIT, however, in being more closely rooted to the human neurophysiological data. Consequently, in GNW, which is derived from the global workspace theory of Baars [10], domain-specific local processing is the source of conscious contents, which only enter provide conscious access when they are integrated into a global workspace, largely involving the PPN. Under this theory, approximately 300 ms is needed to build up an 'ignition' of widespread conscious activity in these cortical areas, in an all or none fashion, with long-range cortical synchronisation occurring in the high gamma band [28, 77].

IIT is a more ambitious, and more controversial, theory that more generally relates network architecture and activity to consciousness. The network can just as easily be a silicon one, if the architecture is appropriate, as a biological one. IIT doesn't just associate integrated information with quantity of consciousness, but *equates* it to the amount of experience. Integrated information under IIT is mathematically formalised and corresponds to that information specified by an entire system (such as a brain) that is more than the sum of its constituent parts. Each experience is uniquely differentiated from all others, and the amount of differentiation possible in a system is also related to the amount of consciousness it can support. Under the latest version of the theory (3.0), new emphasis has been given to conceptual structure and its role in consciousness [76, 73]. Although the specific details of IIT are disputed [78, 79], and its mathematical formalism is unsuitable for practical calculations, many researchers are sympathetic to its strong emphasis on integrated information as a hallmark of consciousness. This view sits well with the behavioural links between consciousness and complex processing, as well as the close association between consciousness and the PPN, given that the PPN is associated with high-level information-based tasks, such as chunking [71, 11].

2.5 Theory and Practice in Imaging Disorders of Consciousness

Active, task-based, DoC imaging paradigms have pioneered research-based assessment of DoC patients, where behavioural clinical assessment alone has proven very unreliable. For instance, Monti and colleagues showed that a patient previously diagnosed as VS could not only perform mental imagery tasks but could also use such tasks to communicate, via fMRI-based brain activity [80, 81]. However, such active tasks rely on the willing participation of the patient, which cannot always be guaranteed in such an ill patient group. Passive paradigms, where the patient has no task to perform, sidestep this issue and may ultimately be more useful clinically. Some of the most promising latest developments in using passive imaging paradigms to assess DoC patients have arisen out of a strong association between consciousness and complex information processing and have been inspired by information-based consciousness theories, such as IIT [5]. For instance, Chennu and colleagues analysed EEG signals in DoC patients and controls using graph theory and found that the patients' networks were impaired at long-range integrated information [82]. Another recent study, based directly on IIT, examined the EEG response to TMS pulses and converted this response signature to a measure of informational complexity, a 'perturbational complexity index' (PCI), by analysing how compressible the EEG signal was [5]. This PCI measure was able to discriminate, at the single subject level, between normal wakefulness, sleep, and general anaesthesia, as well as between different DoC patient groups.

2.6 Conclusions

Thanks in large part to the advent of a range of sophisticated neuroimaging techniques, the scientific study of consciousness has made considerable progress over the last two decades, and there is now much agreement about the neural correlates of consciousness, especially involving the PPN, its role in complex cognition, and its close association with integrated information. This is leading to exciting new techniques by which to diagnose DoC patients, predict recovery, and ultimately index residual conscious level, not just in a binary present or absent way, but along a continuum of awareness (such as the PCI index). DoC research is not only useful for its clinical potential but as a test-bed for empirical results and theoretical stances. Future work needs to further unpick the network and informational mechanics of consciousness, so that we can move beyond mere correlates of consciousness and potentially provide a more nuanced understanding of what impairments DoC patients suffer from and how they can best be helped.

References

1. Bor D (2012) The ravenous brain: How the new science of consciousness explains our insatiable search for meaning. Basic Books, New York
2. Seth AK (2010) The grand challenge of consciousness. Front Psychol 1:5. doi:10.3389/fpsyg.2010.00005
3. Zeman A (2005) What in the world is consciousness? Prog Brain Res 150:1–10. doi:10.1016/S0079-6123(05)50001-3, S0079-6123(05)50001-3 [pii]
4. Tononi G, Koch C (2008) The neural correlates of consciousness: an update. Ann N Y Acad Sci 1124:239–261. doi:10.1196/annals.1440.004, 1124/1/239 [pii]
5. Casali AG, Gosseries O, Rosanova M et al (2013) A theoretically based index of consciousness independent of sensory processing and behavior. Sci Transl Med 5:198ra105. doi:10.1126/scitranslmed.3006294
6. Broadbent DE (1958) Perception and communication. Pergamon Press, London
7. Baddeley AD, Hitch G (1974) Working memory. In: Bower GA (ed) Recent advances in learning and motivation. Academic Press, New York
8. Norman DA, Shallice T (1986) Willed and automatic control of behavior. In: Davidson RJ, Schwarts GE, Shapiro D (eds) Consciousness and self-regulation. Advances in research and theory. Plenum Press, New York, pp 1–18
9. Baars BJ (1988) A cognitive theory of consciousness. Cambridge University Press, New York
10. Baars BJ (2005) Global workspace theory of consciousness: toward a cognitive neuroscience of human experience. Prog Brain Res 150:45–53. doi:10.1016/S0079-6123(05)50004-9, S0079-6123(05)50004-9 [pii]
11. Bor D, Seth AK (2012) Consciousness and the prefrontal parietal network: Insights from attention, working memory, and chunking. Front Psychol 3:1–14. doi:10.3389/fpsyg.2012.00063
12. Koch C, Tsuchiya N (2007) Attention and consciousness: two distinct brain processes. Trends Cogn Sci 11:16–22. doi:10.1016/j.tics.2006.10.012
13. Van Boxtel JJ, Tsuchiya N, Koch C (2010) Opposing effects of attention and consciousness on afterimages. Proc Natl Acad Sci U S A 107:8883–8888. doi:10.1073/pnas.0913292107, 0913292107 [pii]
14. Van Boxtel JJA, Tsuchiya N, Koch C (2010) Consciousness and attention: on sufficiency and necessity. Front Psychol 2:12. doi:10.3389/fpsyg.2010.00217
15. Posner MI (1994) Attention: the mechanisms of consciousness. Proc Natl Acad Sci U S A 91:7398–7403. doi:10.1073/pnas.91.16.7398
16. Rees G, Lavie N (2001) What can functional imaging reveal about the role of attention in visual awareness? Neuropsychologia 39:1343–1353. doi:10.1016/s0028-3932(01)00122-1
17. Bressan P, Pizzighello S (2008) The attentional cost of inattentional blindness. Cognition 106:370–383. doi:10.1016/j.cognition.2007.03.001, S0010-0277(07)00080-7 [pii]
18. Van Den Bussche E, Hughes G, Van Humbeeck N, Reynvoet B (2010) The relation between consciousness and attention: an empirical study using the priming paradigm. Conscious Cogn 19:86–97. doi:10.1016/j.concog.2009.12.019
19. Kentridge RW, Nijboer TC, Heywood CA (2008) Attended but unseen: visual attention is not sufficient for visual awareness. Neuropsychologia 46:864–869. doi:10.1016/j.neuropsychologia.2007.11.036, S0028-3932(07)00417-4 [pii]
20. Norman LJ, Heywood CA, Kentridge RW (2013) Object-based attention without awareness. Psychol Sci 24:836–843. doi:10.1177/0956797612461449

21. Tapia E, Breitmeyer BG, Shooner CR (2010) Role of task-directed attention in nonconscious and conscious response priming by form and color. J Exp Psychol Hum Percept Perform 36:74–87. doi:10.1037/a0017166, 2010-01340-017 [pii]
22. Carrasco M, Ling S, Read S (2004) Attention alters appearance. Nat Neurosci 7:308–313. doi:10.1038/nn1194, nn1194 [pii]
23. Carrasco M, Penpeci-Talgar C, Eckstein M (2000) Spatial covert attention increases contrast sensitivity across the CSF: support for signal enhancement. Vision Res 40:1203–1215, doi: S0042-6989(00)00024-9 [pii]
24. Thorpe S, Fize D, Marlot C (1996) Speed of processing in the human visual system. Nature 381:520–522. doi:10.1038/381520a0
25. Chelazzi L, Duncan J, Miller EK, Desimone R (1998) Responses of neurons in inferior temporal cortex during memory-guided visual search. J Neurophysiol 80:2918–2940
26. Cohen MA, Alvarez GA, Nakayama K (2011) Natural-scene perception requires attention. Psychol Sci 22:1165–1172. doi:10.1177/0956797611419168
27. Cohen MA, Cavanagh P, Chun MM, Nakayama K (2012) The attentional requirements of consciousness. Trends Cogn Sci 16:411–417. doi:10.1016/j.tics.2012.06.013
28. Dehaene S, Changeux JP (2011) Experimental and theoretical approaches to conscious processing. Neuron 70:200–227. doi:10.1016/j.neuron.2011.03.018
29. Mack A, Rock I (1998) Inattentional blindness. MIT Press, Cambridge, MA
30. Simons DJ, Chabris CF (1999) Gorillas in our midst: sustained inattentional blindness for dynamic events. Perception 28:1059–1074
31. Rensink RA, O'Regan JK, Clark JJ, Clark' JJ (1997) To see or not to see: the need for attention to perceive changes in scenes. Psychol Sci 8:368–373. doi:10.1111/j.1467-9280.1997.tb00427.x
32. Cowan N (2001) The magical number 4 in short-term memory: a reconsideration of mental storage capacity. Behav Brain Sci 24:85–87
33. Husain M, Kennard C (1996) Visual neglect associated with frontal lobe infarction. J Neurol 243:652–657
34. Parton A, Malhotra P, Husain M (2004) Hemispatial neglect. J Neurol Neurosurg Psychiatry 75:13–21
35. Sarri M, Blankenburg F, Driver J (2006) Neural correlates of crossmodal visual-tactile extinction and of tactile awareness revealed by fMRI in a right-hemisphere stroke patient. Neuropsychologia 44:2398–2410
36. Dehaene S, Kerszberg M, Changeux JP (1998) A neuronal model of a global workspace in effortful cognitive tasks. Proc Natl Acad Sci U S A 95:14529–14534. doi:10.1073/pnas.95.24.14529
37. Baumeister RF, Masicampo EJ (2010) Conscious thought is for facilitating social and cultural interactions: how mental simulations serve the animal-culture interface. Psychol Rev 117:945–971. doi:10.1037/a0019393, 2010-14834-009 [pii]
38. Melloni L, Schwiedrzik CM, Muller N et al (2011) Expectations change the signatures and timing of electrophysiological correlates of perceptual awareness. J Neurosci 31:1386–1396. doi:10.1523/JNEUROSCI.4570-10.2011, 31/4/1386 [pii]
39. Summerfield C, Koechlin E (2008) A neural representation of prior information during perceptual inference. Neuron 59:336–347. doi:10.1016/j.neuron.2008.05.021
40. Logothetis NK (1998) Single units and conscious vision. Philos Trans R Soc Lond B Biol Sci 353:1801–1818. doi:10.1098/rstb.1998.0333
41. Logothetis NK, Leopold DA, Sheinberg DL (1996) What is rivalling during binocular rivalry? Nature 380:621–624
42. Panagiotaropoulos TI, Deco G, Kapoor V, Logothetis NK (2012) Neuronal discharges and gamma oscillations explicitly reflect visual consciousness in the lateral prefrontal cortex. Neuron 74:924–935. doi:10.1016/j.neuron.2012.04.013
43. Lumer ED, Friston KJ, Rees G (1998) Neural correlates of perceptual rivalry in the human brain. Science (80-) 280:1930–1934
44. Rees G (2007) Neural correlates of the contents of visual awareness in humans. Philos Trans R Soc Lond B Biol Sci 362:877–886. doi:10.1098/rstb.2007.2094, P6N2252036135762 [pii]

45. Dehaene S, Naccache L, Cohen L et al (2001) Cerebral mechanisms of word masking and unconscious repetition priming. Nat Neurosci 4:752–758. doi:10.1038/89551 89551 [pii]
46. Sadaghiani S, Hesselmann G, Kleinschmidt A (2009) Distributed and antagonistic contributions of ongoing activity fluctuations to auditory stimulus detection. J Neurosci 29:13410–13417. doi:10.1523/JNEUROSCI.2592-09.2009, 29/42/13410 [pii]
47. Boly M, Balteau E, Schnakers C et al (2007) Baseline brain activity fluctuations predict somatosensory perception in humans. Proc Natl Acad Sci U S A 104:12187–12192. doi:10.1073/pnas.0611404104
48. Lau HC, Passingham RE (2006) Relative blindsight in normal observers and the neural correlate of visual consciousness. Proc Natl Acad Sci U S A 103:18763–18768. doi:10.1073/pnas.0607716103
49. Del Cul A, Dehaene S, Reyes P et al (2009) Causal role of prefrontal cortex in the threshold for access to consciousness. Brain 132:2531–2540. doi:10.1093/brain/awp111, awp111 [pii]
50. Knight RT, Grabowecky M (1995) Escape from linear time: prefrontal cortex and conscious experience. In: Gazzaniga M (ed) The new cognitive neurosciences. MIT Press, Massachusetts, pp 1357–1371
51. Simons JS, Peers PV, Mazuz YS et al (2010) Dissociation between memory accuracy and memory confidence following bilateral parietal lesions. Cereb Cortex 20:479–485. doi:10.1093/cercor/bhp116, bhp116 [pii]
52. Beck DM, Muggleton N, Walsh V, Lavie N (2006) Right parietal cortex plays a critical role in change blindness. Cereb Cortex 16:712–717. doi:10.1093/cercor/bhj017
53. Turatto M, Sandrini M, Miniussi C (2004) The role of the right dorsolateral prefrontal cortex in visual change awareness. Neuroreport 15:2549–2552
54. Rounis E, Maniscalco B, Rothwell JC et al (2010) Theta-burst transcranial magnetic stimulation to the prefrontal cortex impairs metacognitive visual awareness. Cogn Neurosci 1:165–175
55. Fleming SM, Weil RS, Nagy Z et al (2010) Relating introspective accuracy to individual differences in brain structure. Science 329:1541–1543. doi:10.1126/science.1191883, 329/5998/1541 [pii]
56. Fleming SM, Huijgen J, Dolan RJ (2012) Prefrontal contributions to metacognition in perceptual decision making. J Neurosci 32:6117–6125. doi:10.1523/JNEUROSCI.6489-11.2012, 32/18/6117 [pii]
57. Kanai R, Bahrami B, Rees G (2010) Human parietal cortex structure predicts individual differences in perceptual rivalry. Curr Biol 20:1626–1630. doi:10.1016/j.cub.2010.07.027
58. Tshibanda L, Vanhaudenhuyse A, Galanaud D et al (2009) Magnetic resonance spectroscopy and diffusion tensor imaging in coma survivors: promises and pitfalls. Prog Brain Res 177:215–229. doi:10.1016/S0079-6123(09)17715-4, S0079-6123(09)17715-4 [pii]
59. Boly M, Garrido MI, Gosseries O et al (2011) Preserved feedforward but impaired top-down processes in the vegetative state. Science (80-) 332:858–862. doi:10.1126/science.1202043
60. Laureys S, Faymonville ME, Luxen A et al (2000) Restoration of thalamocortical connectivity after recovery from persistent vegetative state. Lancet 355:1790–1791, doi: S0140673600022716 [pii]
61. Maquet P, Degueldre C, Delfiore G et al (1997) Functional neuroanatomy of human slow wave sleep. J Neurosci 17:2807–2812
62. Davis MH, Coleman MR, Absalom AR et al (2007) Dissociating speech perception and comprehension at reduced levels of awareness. Proc Natl Acad Sci U S A 104:16032–16037
63. Kaisti KK, Metsähonkala L, Teräs M et al (2002) Effects of surgical levels of propofol and sevoflurane anesthesia on cerebral blood flow in healthy subjects studied with positron emission tomography. Anesthesiology 96:1358–1370. doi:10.1097/00000542-200206000-00015
64. Laureys S, Owen AM, Schiff ND (2004) Brain function in coma, vegetative state, and related disorders. Lancet Neurol 3:537–546. doi:10.1016/S1474-4422(04)00852-X
65. Monti MM, Laureys S, Owen AM (2010) The vegetative state. BMJ 341:c3765. doi:10.1136/bmj.c3765
66. Fernández-Espejo D, Bekinschtein T, Monti MM et al (2011) Diffusion weighted imaging distinguishes the vegetative state from the minimally conscious state. Neuroimage 54:103–112. doi:10.1016/j.neuroimage.2010.08.035

67. Schiff ND, Giacino JT, Kalmar K et al (2007) Behavioural improvements with thalamic stimulation after severe traumatic brain injury. Nature 448:600–603. doi:10.1038/nature06818
68. Duncan J (2006) Brain mechanisms of attention. Q J Exp Psychol 59:2–27
69. Bor D, Cumming N, Scott CEL, Owen AM (2004) Prefrontal cortical involvement in verbal encoding strategies. Eur J Neurosci 19:3365–3370. doi:10.1111/j.1460-9568.2004.03438.x
70. Bor D, Duncan J, Wiseman RJ, Owen AM (2003) Encoding strategies dissociate prefrontal activity from working memory demand. Neuron 37:361–367. doi:10.1016/S0896-6273(02)01171-6
71. Bor D, Owen AM (2007) A common prefrontal-parietal network for mnemonic and mathematical recoding strategies within working memory. Cereb Cortex 17:778–786. doi:10.1093/cercor/bhk035
72. Lamme VA (2010) How neuroscience will change our view on consciousness. Cogn Neurosci 1:204–220. doi:10.1080/17588921003731586
73. Tononi G, Koch C (2015) Consciousness: here, there and everywhere? Phil Trans R Soc B 370: 20140167. doi:10.1098/rstb.2014.0167
74. Modha DS, Singh R (2010) Network architecture of the long-distance pathways in the macaque brain. Proc Natl Acad Sci U S A 107:13485–13490. doi:10.1073/pnas.1008054107, 1008054107 [pii]
75. Barttfeld P, Uhrig L, Sitt JD et al (2015) Signature of consciousness in the dynamics of resting-state brain activity. Proc Natl Acad Sci 112:201418031. doi:10.1073/pnas.1418031112
76. Oizumi M, Albantakis L, Tononi G (2014) From the phenomenology to the mechanisms of consciousness: Integrated Information Theory 3.0. PLoS Comput Biol 10, e1003588. doi:10.1371/journal.pcbi.1003588
77. Gaillard R, Dehaene S, Adam C et al (2009) Converging intracranial markers of conscious access. PLoS Biol 7, e61
78. Aaronson S (2014) Why i am not an integrated information theorist (or, The unconscious expander). In: Shtetl Optim. Blog Scott Aaronson. http://www.scottaaronson.com/blog/?p=1799
79. Barrett AB, Seth AK (2011) Practical measures of integrated information for time-series data. PLoS Comput Biol 7:e1001052. doi:10.1371/journal.pcbi.1001052
80. Monti MM, Vanhaudenhuyse A, Coleman MR et al (2010) Willful modulation of brain activity in disorders of consciousness. N Engl J Med 362:579–589. doi:10.1056/NEJMoa0905370, NEJMoa0905370 [pii]
81. Owen AM, Coleman MR, Boly M et al (2006) Detecting awareness in the vegetative state. Science (80-) 313:1402. doi:10.1126/science.1130197, 313/5792/1402 [pii]
82. Chennu S, Finoia P, Kamau E et al (2014) Spectral signatures of reorganised brain networks in disorders of consciousness. PLoS Comput Biol. doi:10.1371/journal.pcbi.1003887

Chapter 3
Behavioral Responsiveness in Patients with Disorders of Consciousness

Lizette Heine, Steven Laureys, and Caroline Schnakers

Abstract Until now, the main way to assess whether a patient is conscious or not is to observe what he can do spontaneously or in response to stimulation. Although there is a growing body of research on the subject, detecting oriented/voluntary responses is still extremely challenging. Motor, verbal, and cognitive impairments; fluctuations of vigilance; and medications with impact on the central nervous system are among the factors complicating the diagnosis. Establishing a proper diagnosis is nevertheless of high clinical relevance when considering patients' prognosis and treatment. In this review, we will characterize the behavioral patterns of the various levels of consciousness, we will explain how challenging it is to detect signs of consciousness, and which tools currently exist to help in the assessment of those signs. Secondarily, we will present preliminary data investigating the interest of various sensory modalities in determining the diagnosis of patients with severe brain injury.

Keywords Vegetative state • Minimally conscious state • Consciousness • Assessment • Diagnosis • Sensory stimulation

3.1 Introduction

Some patients surviving extensive brain damage only regain limited levels of consciousness. Until now, the main way to assess whether a patient is conscious or not is to observe what he/she can do spontaneously or in response to stimulation. Although there is a growing body of research on the subject, detecting oriented/voluntary

L. Heine, MSc
Coma Science Group, Cyclotron Research Centre, University of Liège, Liège, Belgium

S. Laureys, MD, PhD
Coma Science Group, Neurology Department, Cyclotron Research Centre, University of Liège, Liège, Belgium

C. Schnakers, PhD (✉)
Department of Neurosurgery, Geffen School of Medicine at University of California Los Angeles, Brain Injury Research Center, Los Angeles, CA, USA
e-mail: cschnakers@ucla.edu

M.M. Monti, W.G. Sannita (eds.), *Brain Function and Responsiveness in Disorders of Consciousness*, DOI 10.1007/978-3-319-21425-2_3

responses is still extremely challenging. Motor, verbal, and cognitive impairments; fluctuations of vigilance; and medications with impact on the central nervous system are among the factors complicating the diagnosis. Establishing a proper diagnosis is nevertheless of high clinical relevance. Conscious and unconscious patients have different outcomes. Patients in a minimally conscious state have a better long-term prognosis compared to those in a vegetative state [1]. Twelve months after brain injury, about half of the patients in minimally conscious state improve and show a good functional outcome, whereas only a very small percentage (3 %) of patients in vegetative state do so [2]. The diagnosis also has an impact on the patients' daily care and therapeutic choices when it comes to the administration of pharmacological interventions such as pain medication or new non-pharmacological interventions such as neurostimulation (deep brain stimulation or transcranial direct current stimulation) [3, 4]. Finally, regarding end-of-life decisions, previous legal cases in several countries have established the right of the medical team to withdraw artificial nutrition and hydration in patients diagnosed as being in a vegetative state [5]. In such context, a correct diagnosis is therefore crucial. In this review, we will characterize the behavioral pattern of the various levels of consciousness, we will explain how challenging it is to detect signs of consciousness, and which tools currently exist to help in the assessment of those signs. Secondarily, we will also present preliminary data investigating the interest of various sensory modalities in determining the diagnosis of patients with severe brain injury.

3.2 Behavioral Pattern in Disorders of Consciousness

When the patient is in a coma, there is no arousal and no consciousness. During this transient condition, patients' eyes are continuously closed (even following stimulation), autonomic functions are reduced, and respiratory assistance is needed [6] (Table 3.1). Most patients recover from a coma within hours to weeks after injury. However, some patients can recover arousal (i.e., open their eyes spontaneously or in response to stimulation) without being conscious (no oriented/voluntary responses). These patients are in a state called "vegetative state" (VS) [7] (Table 3.1). In this state, breathing occurs without assistance since autonomic functions (e.g., cardiovascular regulation, thermoregulation) are preserved. The patients may also moan, demonstrate smiling, crying, or grimacing even though inappropriate and appearing out of context [7, 8]. This state can be either transient or persistent (when above a month post-injury). After a year for traumatic etiologies and 3 months for nontraumatic etiologies, the VS can be considered as permanent. These patients have, in that case, less than 5 % of chances to recover. Only then, the ethical and legal issues around withdrawal of hydration and nutrition may be discussed [9]. Note that, given the negative connotation of the term "vegetative state," The European Task Force on Disorders of Consciousness has recently proposed to use the more neutral and descriptive term "unresponsive wakefulness syndrome" (VS/UWS) [10].

Table 3.1 Summary of the behavioral features for coma, VS/UWS, MCS−, MCS+, and emergence from MCS

Level of consciousness	Behavioral features
Coma	No arousal/eye opening
	Impaired spontaneous breathing/brainstem reflexes
	No oriented or purposeful behaviors
	No groans, vocalizations, or verbalizations
	No language comprehension/response to command
Vegetative state/unresponsive wakefulness syndrome	Arousal/spontaneous or stimulus-induced eye opening
	Preserved spontaneous breathing/brainstem reflexes
	No oriented or purposeful behaviors
	Groans and/or vocalizations but no verbalizations
	No language comprehension/response to command
Minimally conscious state	Fluctuation of vigilance (MCS−/+)
	Preserved spontaneous breathing/brainstem reflexes
	MCS−: object localization-reaching-manipulation and/or sustained visual fixation and/or visual pursuit and/or automatic motor behavior and/or localization to pain
	MCS+: command following and/or object recognition and/or intelligible verbalization and/or intentional communication
	Emergence: functional communication and/or functional object use on at least two consecutive assessments

Consciousness recovery consists of regaining fluctuating but reproducible nonreflexive-oriented and/or voluntary behaviors. Such state is called the "minimally conscious state" (MCS) [11] (Table 3.1). Behaviors that suggest consciousness are, for example, command following, visual pursuit, object localization, or contingent responses to emotional stimuli. MCS has recently been divided into two categories, MCS+ (plus) and MCS- (minus), based on the complexity of behavioral responses. Patients in an MCS- show nonreflexive-oriented responses such as visual pursuit or localization to noxious stimuli, while MCS+ refers to patients showing nonreflexive voluntary responses such as command following, intelligible verbalization, and/or nonfunctional communication [12, 13]. When patients demonstrate reliable "functional communication" (i.e., accurate yes-no responses to situational orientation questions) or "functional object use" (i.e., appropriate use of different common objects) on two consecutive assessments, the patient is considered to have emerged from the MCS (EMCS) [11] (Table 3.1). After emerging from MCS, these patients are not considered as being in a disorder of consciousness anymore. However, they often remain confused, disoriented, and sometimes agitated. The term "acute confusional state" (ACS) has recently been used to describe these patients [14].

3.3 Misdiagnosis

Differentiating MCS from VS/UWS can be challenging since voluntary and reflexive behaviors can be difficult to distinguish and subtle signs of consciousness may be missed. The development of diagnostic criteria for MCS [11] would reasonably be expected to reduce the incidence of misdiagnosis relative to the rates reported before these criteria were established [15, 16]. However, recent studies found that around 40 % of patients believed to be in VS/UWS were still misdiagnosed [17, 18] (Fig. 3.1).

The high rate of misdiagnosis likely reflects different sources of variance. Variance in diagnostic accuracy may result from biases contributed by the examiner, the environment, and/or the patient. First, examiner errors may arise when the range of behaviors sampled is too narrow, response-time windows are over- or under-inclusive, criteria for judging purposeful responses are poorly defined, and examinations are conducted too infrequently to capture the full range of behavioral fluctuation. The use of standardized rating scales offers some protection from these errors, although failure to adhere to specific administration and scoring guidelines may jeopardize diagnostic accuracy. Second, the environment in which the patient is evaluated may bias assessment findings. Paralytic and sedative medications, restricted range of movement stemming from restraints and immobilization techniques, poor positioning, and excessive ambient noise/heat/light can decrease or

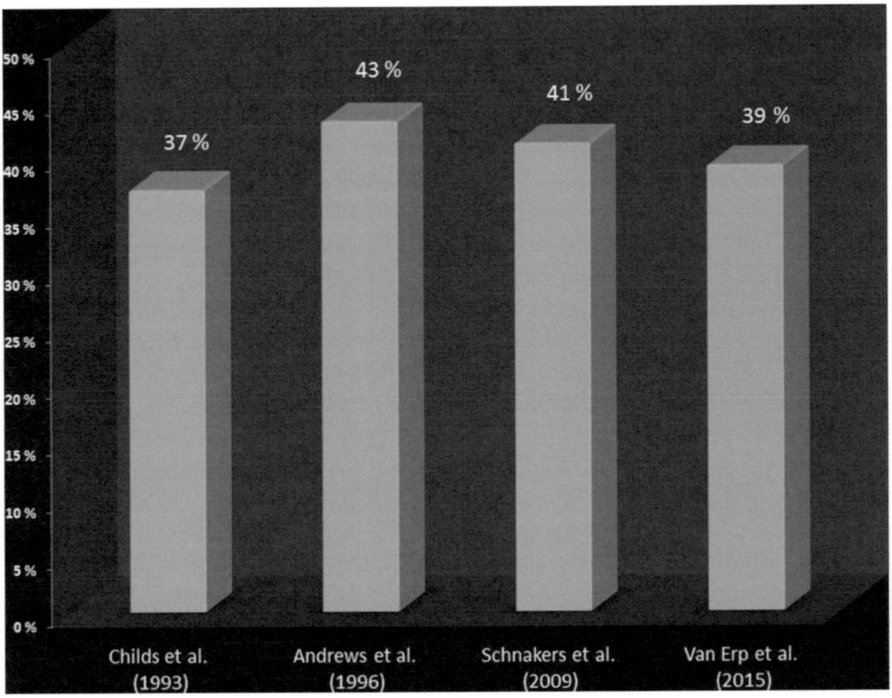

Fig. 3.1 Previous findings on misdiagnosis rate in patients with severe brain injury

distort voluntary behavioral responses. The last source of variance concerns the patient. Fluctuations in arousal level, fatigue, subclinical seizure activity, occult illness, pain, cortical sensory deficits (e.g., cortical blindness/deafness), motor impairment (e.g., generalized hypotonus, spasticity, or paralysis), or cognitive deficits (e.g., aphasia, apraxia, agnosia) constitute a bias to the behavioral assessment and therefore decrease the probability to observe signs of consciousness.

Some sources of error can be avoided, but this is not always possible or within the examiner's control. It is, however, particularly crucial to optimize the way consciousness assessments are performed as clinical management, from treatment of pain to end-of-life decision-making, often depends on behavioral observations. For this reason, the use of standardized and sensitive behavioral scales can substantially help clinicians to detect subtle signs of consciousness.

3.4 Clinical Assessment of Disorders of Consciousness

Behavioral assessment is based on two main components: wakefulness and awareness. Wakefulness refers to the patient's level of arousal and is assessed by observing eye opening. Awareness is related to subjective experiences and can be subdivided into awareness of the external world (i.e., perception of the environment or "consciousness") and awareness of the internal world (i.e., stimulus-independent thoughts such as mental imagery and inner speech or "self-awareness"). Raters assessing patients with severe brain injury will mainly assess consciousness of the environment, since self-awareness is difficult to evaluate when only based on bedside observations and not on patients' report. The assessment of consciousness can be done through repeated examinations revealing reproducible, oriented, or voluntary behavioral responses to various stimuli (the most common being auditory, verbal, and motor stimuli). The first scale widely used and known for assessing severely brain-injured patients recovering from coma is the Glasgow coma scale (GCS) [19]. This scale is short and can easily be incorporated into routine clinical care. Despite its widespread use, the GCS has been criticized for fluctuant inter-rater reliability and problems of scoring in patients with ocular trauma, tracheostomy, or ventilatory support [20]. The Full Outline of UnResponsiveness (FOUR) has been developed to replace the GCS for assessing severely brain-injured patients in intensive care [21]. The scale includes four subscales assessing motor and ocular responses, brainstem reflexes, and breathing. The total score ranges from 0 to 16. Unlike the GCS, the FOUR does not assess verbal functions to accommodate the high number of intubated patients in intensive care. It also assesses brainstem reflexes and breathing and, therefore, helps to better monitor comatose and VS/UWS patients. The FOUR also tracks emergence from VS/UWS since it includes the assessment of early signs of consciousness such as visual pursuit. The scale is globally more sensitive than the GCS for diagnosing MCS but like the GCS is not adapted to a rehabilitation setting.

Since the 1970s, a high number of scales have been validated for being used in subacute and chronic patients with severe brain injury (Table 3.2). Recently, the

Table 3.2 Behavioral responses assessed by scales developed for patients with disorders of consciousness

Name of the scale (Reference)	Response to command	Contingent emotional response	Object localization/ manipulation	Intelligible verbalizations	Oriented response to sensory stimulation				
					V	N	T	O	G
Coma Recovery Scale-Revised [24]	*		*	*	*	*			
Western Neuro Sensory Stimulation Profile [25]	*	*	*	*	*		*	*	
Sensory Modality Assessment & Rehabilitation Technique [34]	*	*	*	*	*		*	*	*
Wessex Head Injury Matrix [23]	*	*	*	*	*				
Disorder of Consciousness Scale [27]	*			*	*		*	*	
Sensory Stimulation Assessment Measure [35]	*		*	*	*		*	*	*
Glasgow Coma Scale [19]	*			*		*			
Reaction Level Scale [36]	*			*	*	*			
Innsbruck Coma Scale [37]						*			
Glasgow-Liège Scale [38]	*			*		*			
Full Outline of UnResponsiveness [21]	*				*	*			
Coma/Near-Coma Scale [39]	*			*	*		*	*	
Comprehensive of Level of Consciousness Scale [40]	*			*	*				

V visual, *N* nociceptive, *T* tactile, *O* olfactory, *G* gustatory, *bold* scale recommended with minor reservations by the ACRM, *italic* scales recommended with moderate reservations by the ACRM

American Congress of Rehabilitation Medicine (ACRM) has conducted a systematic evidence-based review of the available scales to provide recommendations for use according to validity, reliability, outcome prediction, and diagnostic sensitivity [22]. Among the scales evaluated, the Wessex Head Injury Matrix (WHIM) has been recommended with moderate reservations. The WHIM was developed to capture changes in patients in VS/UWS through emergence from post-traumatic amnesia [23]. This tool is particularly sensitive to detect changes in patients in MCS not captured by other scales such as the GCS. The WHIM has been structured according to the sequence of recovery observed in 88 patients recovering from traumatic brain injury. The scale assesses arousal level and concentration, visual pursuit, communication, cognition (i.e., memory and spatiotemporal orientation), and social behaviors. The WHIM score represents the rank of the most complex behavior observed. Despite a good validity, its reliability is still unproven, and, even though superior to the GCS, its diagnostic sensitivity is lower than other standardized scales such as the Coma Recovery Scale-Revised (CRS-R) [24]. In fact, according to the ACRM, the CRS-R is the most reliable tool for differentiating disorders of consciousness and received the strongest recommendation with minor reservations [22]. This scale was developed in 1991 and revised in 2004. Its primary purpose is to differentiate VS/UWS from MCS and MCS from EMCS. It measures auditory, visual, motor, and verbal functions as well as communication and arousal. Each of these subscales is hierarchically structured; the lowest scores reflect reflexive behaviors, while the highest scores indicate cognitively mediated behaviors. This scale has clear definitions for both the administration and the scoring of each item. The CRS-R can be administered reliably by trained examiners and produces reasonably stable scores over repeated assessments. Validity analyses have shown that the CRS-R is capable of discriminating patients in MCS from those in VS/UWS better than the GCS, the FOUR, and the WHIM [24].

Other scales such as the Western Neuro Sensory Stimulation Profile (WNSSP) [25], the Sensory Modality Assessment Technique (SMART) [26], and the Disorders of Consciousness Scale (DOCS) [27] have acceptable standardized administration and scoring procedures and have also been recommended with moderate reservations by the ACRM. On the contrary to the CRS-R whose main purpose is the diagnosis, the WNSSP, the SMART, and the DOCS are rather used when applying a sensory stimulation treatment to patients with severe brain injury. Sensory stimulation programs usually consist in presenting different types of environmental stimuli to the patient in order to optimize her/his consciousness level. These programs are supposed to constitute enriched environments which are supposed to enhance synaptic reinnervation, improve brain plasticity, and therefore accelerate the recovery from coma. However, even though numerous studies investigated the interest of these sensory stimulation programs, none of these studies has proven the efficacy of such treatment since the findings did not allow to differentiate spontaneous recovery from recovery due to treatment. Despite this, scales such as the WNSSP, the SMART, or the DOCS could still be interesting in a diagnostic context since they include the assessment of more sensory modalities than the CRS-R (i.e., tactile, olfactory, and gustatory modalities). The interest of those modalities for detecting signs of consciousness has nevertheless never been evaluated.

3.5 Can More Sensory Modalities Increase Diagnostic Sensitivity?

It has previously been shown that some sensory modalities are more sensitive to detect consciousness than others. In studies investigating misdiagnosis, oriented eye movements (i.e., visual pursuit and fixation) have been reported as the responses the most frequently missed during behavioral assessments [28, 17, 18]. In parallel, the visual modality of the CRS-R has been shown as the subscale allowing the highest detection of MCS as compared to the auditory, motor, or verbal modalities [29, 30]. Oriented visual responses are particularly interesting to detect since it is one of the first signs of consciousness appearing during patients' recovery and as it is associated with good outcome [31, 32, 2]. Until now, no study has investigated the interest of other sensory modalities (such as tactile, olfactory, and gustatory) when assessing consciousness, even though several scales recommended by the ACRM include such modalities (Table 3.2).

In a preliminary study, we therefore decided to investigate the interest of tactile, olfactory, and gustatory modalities in the assessment of consciousness. We assessed 38 patients (46 ± 16 years old, 17 traumatic, 21 chronic) diagnosed as being in a VS/UWS ($n = 15$) or in a MCS ($n = 23$) by using the CRS-R. Tactile, olfactory, and gustatory stimuli used in the WNSSP, the SMART, and the DOCS have been administered in each patient in a randomized order. Tactile stimuli included tap on the shoulder, nasal swab, feather (applied on arms, fingers, and face), air into the neck, hair touching, vibration on the arm, scrub (i.e., kitchen scouring pad applied over the arm), and firm hand pressure on the arm. Each of these stimuli was applied for 10 s on both sides of the body on three consecutive trials. Olfactory stimuli included vinegar, syrup, and ammonia which were held under the patients' nose for 10 s (patient's mouth closed) on three consecutive trials. In case of tracheotomy, the entrance of the cannula was covered. Gustatory stimuli included vinegar and syrup. A stick soaked of this flavor was introduced into the patient's mouth for 10 s on three consecutive trials. Several recommendations had to be followed such as applying the treatment while the patients were in a wakeful state with eyes open in a setting with minimal ambient noise and respecting a 30 min rest before each session (i.e., absence of nursing care). Oriented responses (e.g., eyes/head toward or away from the stimulus, hand toward or pushes away the stimulus, congruent facial expression, mouth opening, or tongue pumping) were considered as present when it was clear and reproducible, meaning it was observed at least two times to exclude reflexive behaviors. The oriented responses obtained using those tactile, olfactory, and gustatory stimulations have then been compared to the diagnosis obtained using the CRS-R. Patients' outcome has also been collected at 1 year after assessment ($n = 27$), using the Glasgow Outcome Scale (GOS) [33].

According to our results (Fig. 3.2), a minority of patients diagnosed as being in a VS/UWS by using the CRS-R showed oriented olfactory or gustatory responses (7 % and 14 %, respectively). The patient for whom we had outcome data (one missing data) did not recover consciousness a year after assessment. Additionally, oriented

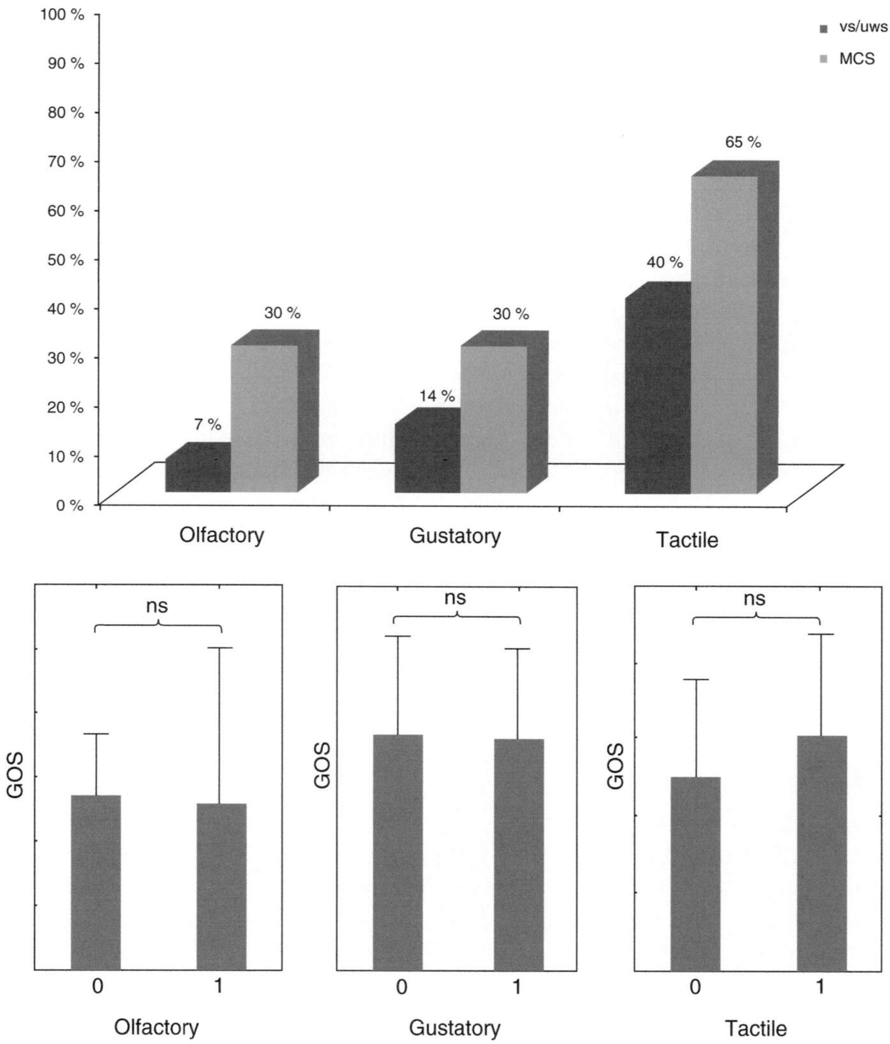

Fig. 3.2 Percentage of oriented responses in VS/UWS and MCS patients (panel **a**) and outcome at 1 year (averaged Glasgow Outcome Score – GOS, with 95 % confidence intervals) according to the absence (*0*) or presence (*1*) of oriented responses (panel **b**) ("ns" indicates difference is nonsignificant ($p>0.05$))

olfactory or gustatory responses were absent in a majority of patients diagnosed as being in a MCS by using the CRS-R (70 %) and in a majority of patients who showed oriented eye movements (61 %). Using tactile stimuli, a higher percentage of patients diagnosed as being in a VS/UWS showed oriented responses (40 %). Oriented tactile responses were present in a majority of patients diagnosed as being in an MCS by using the CRS-R (65 %) and in a majority of patients who showed oriented eye movements (83 %). When considering the stimulus leading to the most frequent oriented responses, the nasal swab helped to detect 80 % of the oriented tactile responses. However, only one of the VS/UWS patients showing oriented tactile responses recovered consciousness a year after assessment (17 %). The patient (50 years old, 50 days after nontraumatic injury) was able to localize a tactile stimulus using her hand. Repeated CRS-R assessments, at that time, showed only reflexive behaviors (i.e., auditory startle, blinking to threat, flexion to noxious stimulation, oral reflexive movements, and arousal with stimulation). Two years after our assessment, the CRS-R indicated an EMCS. Finally, to test whether the outcome measured by the GOS differs according to the presence or absence of an oriented response, *U* Mann-Whitney tests were performed. There was no statistical difference for olfactory ($U=51.5$; $p=0.61$), gustatory ($U=49$; $p=0.5$), and tactile ($U=76.5$; $p=0.51$) modalities.

Considering our data, oriented olfactory and gustatory responses do not seem to be linked to consciousness since they are not observed in the majority of significant proportion of conscious patients and since they are not associated with consciousness recovery. Oriented tactile responses seem to be observed in most conscious patients but are not clearly related to consciousness recovery and could be false positives. This preliminary study hence seems to indicate that adding sensory modalities such as olfactory, gustatory, or tactile modalities to the CRS-R does not constitute a further help for decreasing the level of misdiagnosis in patients with disorders of consciousness.

3.6 Conclusion

Establishing a proper diagnosis is very important in the care of patients with severe brain injury. However, clinical assessment is difficult and can often lead to a misdiagnosis of the level of consciousness. The use of sensitive standardized tools is therefore crucial when establishing the diagnosis. The CRS-R is currently the most reliable and valid scale available and constitutes a substantial help in the differentiation of conscious vs. unconscious patients. Finally, even though our findings need to be replicated in a bigger sample, using gustatory, olfactory, or tactile stimuli that are included in several behavioral scales for the assessment of disorders of consciousness do not seem to be of further help when detecting consciousness in patients with severe brain injury.

References

1. Luaute J, Maucort-Boulch D, Tell L, Quelard F, Sarraf T, Iwaz J et al (2010) Long-term outcomes of chronic minimally conscious and vegetative states. Neurology 75(3):246–252. doi:10.1212/WNL.0b013e3181e8e8df
2. Giacino JT, Kezmarsky MA, DeLuca J, Cicerone KD (1991) Monitoring rate of recovery to predict outcome in minimally responsive patients. Arch Phys Med Rehabil 72(11):897–901

3. Schiff ND, Giacino JT, Kalmar K, Victor JD, Baker K, Gerber M et al (2007) Behavioural improvements with thalamic stimulation after severe traumatic brain injury. Nature 448(7153): 600–603. doi:10.1038/nature06041
4. Thibaut A, Bruno MA, Ledoux D, Demertzi A, Laureys S (2014) tDCS in patients with disorders of consciousness: sham-controlled randomized double-blind study. Neurology 82(13):1112–1118. doi:10.1212/WNL.0000000000000260
5. Racine E, Rodrigue C, Bernat JL, Riopelle R, Shemie SD (2010) Observations on the ethical and social aspects of disorders of consciousness. Can J Neurol Sci 37(6):758–768
6. Posner JB, Saper JB, Schiff ND, Plum F (2007) Plum and Posner's diagnosis of stupor and coma, 4th edn, Contemporary neurology series. Oxford University Press, Oxford/New York
7. Jennett B, Plum F (1972) Persistent vegetative state after brain damage. RN 35(10):ICU1–ICU4
8. The Multi-Society Task Force on PVS (1994) Medical aspects of the persistent vegetative state (1). N Engl J Med 330(21):1499–1508. doi:10.1056/NEJM199405263302107
9. The Multi-Society Task Force on PVS (1994) Medical aspects of the persistent vegetative state (2). N Engl J Med 330(22):1572–1579. doi:10.1056/NEJM199406023302206
10. Laureys S, Celesia GG, Cohadon F, Lavrijsen J, Leon-Carrion J, Sannita WG et al (2010) Unresponsive wakefulness syndrome: a new name for the vegetative state or apallic syndrome. BMC Med 8:68. doi:10.1186/1741-7015-8-68
11. Giacino JT, Ashwal S, Childs N, Cranford R, Jennett B, Katz DI et al (2002) The minimally conscious state: definition and diagnostic criteria. Neurology 58(3):349–353
12. Bruno MA, Majerus S, Boly M, Vanhaudenhuyse A, Schnakers C, Gosseries O et al (2012) Functional neuroanatomy underlying the clinical subcategorization of minimally conscious state patients. J Neurol 259(6):1087–1098. doi:10.1007/s00415-011-6303-7
13. Bruno MA, Vanhaudenhuyse A, Thibaut A, Moonen G, Laureys S (2011) From unresponsive wakefulness to minimally conscious PLUS and functional locked-in syndromes: recent advances in our understanding of disorders of consciousness. J Neurol 258(7):1373–1384. doi:10.1007/s00415-011-6114-x
14. Sherer M, Nakase-Thompson R, Yablon SA, Gontkovsky ST (2005) Multidimensional assessment of acute confusion after traumatic brain injury. Arch Phys Med Rehabil 86(5):896–904. doi:10.1016/j.apmr.2004.09.029
15. Andrews K, Murphy L, Munday R, Littlewood C (1996) Misdiagnosis of the vegetative state: retrospective study in a rehabilitation unit. BMJ 313(7048):13–16
16. Childs NL, Mercer WN, Childs HW (1993) Accuracy of diagnosis of persistent vegetative state. Neurology 43(8):1465–1467
17. Schnakers C, Vanhaudenhuyse A, Giacino J, Ventura M, Boly M, Majerus S et al (2009) Diagnostic accuracy of the vegetative and minimally conscious state: clinical consensus versus standardized neurobehavioral assessment. BMC Neurol 9:35. doi:10.1186/1471-2377-9-35
18. van Erp WS, Lavrijsen JC, Vos PE, Bor H, Laureys S, Koopmans RT (2015) The vegetative state: prevalence, misdiagnosis, and treatment limitations. J Am Med Dir Assoc 16(1):85 e9–e14. doi:10.1016/j.jamda.2014.10.014
19. Teasdale G, Jennett B (1974) Assessment of coma and impaired consciousness. A practical scale. Lancet 2(7872):81–84
20. Laureys S, Bodart O, Gosseries O (2014) The Glasgow Coma Scale: time for critical reappraisal? Lancet Neurol 13(8):755–757. doi:10.1016/S1474-4422(14)70152-8
21. Wijdicks EF, Bamlet WR, Maramattom BV, Manno EM, McClelland RL (2005) Validation of a new coma scale: The FOUR score. Ann Neurol 58(4):585–593. doi:10.1002/ana.20611
22. Seel RT, Sherer M, Whyte J, Katz DI, Giacino JT, Rosenbaum AM et al (2010) Assessment scales for disorders of consciousness: evidence-based recommendations for clinical practice and research. Arch Phys Med Rehabil 91(12):1795–1813. doi:10.1016/j.apmr.2010.07.218
23. Shiel A, Horn SA, Wilson BA, Watson MJ, Campbell MJ, McLellan DL (2000) The Wessex Head Injury Matrix (WHIM) main scale: a preliminary report on a scale to assess and monitor patient recovery after severe head injury. Clin Rehabil 14(4):408–416
24. Giacino JT, Kalmar K, Whyte J (2004) The JFK Coma Recovery Scale-Revised: measurement characteristics and diagnostic utility. Arch Phys Med Rehabil 85(12):2020–2029

25. Ansell BJ, Keenan JE (1989) The Western Neuro Sensory Stimulation Profile: a tool for assessing slow-to-recover head-injured patients. Arch Phys Med Rehabil 70(2):104–108
26. Gill-Thwaites H, Munday R (2004) The Sensory Modality Assessment and Rehabilitation Technique (SMART): a valid and reliable assessment for vegetative state and minimally conscious state patients. Brain Inj 18(12):1255–1269
27. Pape TL, Senno RG, Guernon A, Kelly JP (2005) A measure of neurobehavioral functioning after coma. Part II: Clinical and scientific implementation. J Rehabil Res Dev 42(1):19–27
28. Childs NL, Mercer WN (1996) Misdiagnosing the persistent vegetative state. Misdiagnosis certainly occurs. BMJ 313(7062):944
29. Vanhaudenhuyse A, Schnakers C, Bredart S, Laureys S (2008) Assessment of visual pursuit in post-comatose states: use a mirror. J Neurol Neurosurg Psychiatry 79(2):223. doi:10.1136/jnnp.2007.121624
30. Schnakers C, Majerus S, Giacino J, Vanhaudenhuyse A, Bruno MA, Boly M et al (2008) A French validation study of the Coma Recovery Scale-Revised (CRS-R). Brain Inj 22(10):786–792. doi:10.1080/02699050802403557
31. Candelieri A, Cortese MD, Dolce G, Riganello F, Sannita WG (2011) Visual pursuit: within-day variability in the severe disorder of consciousness. J Neurotrauma 28(10):2013–2017. doi:10.1089/neu.2011.1885
32. Dolce G, Lucca LF, Candelieri A, Rogano S, Pignolo L, Sannita WG (2011) Visual pursuit in the severe disorder of consciousness. J Neurotrauma 28(7):1149–1154. doi:10.1089/neu.2010.1405
33. Teasdale GM, Pettigrew LE, Wilson JT, Murray G, Jennett B (1998) Analyzing outcome of treatment of severe head injury: a review and update on advancing the use of the Glasgow Outcome Scale. J Neurotrauma 15(8):587–597
34. Gill-Thwaites H (1997) The Sensory Modality Assessment Rehabilitation Technique–a tool for assessment and treatment of patients with severe brain injury in a vegetative state. Brain Inj 11(10):723–34
35. Rader MA, Ellis DW (1994) The Sensory Stimulation Assessment Measure (SSAM): a tool for early evaluation of severely brain-injured patients. Brain Inj 8(4):309–21
36. Starmark JE, Stålhammar D, Holmgren E (1988) The Reaction Level Scale (RLS85). Manual and guidelines. Acta Neurochir (Wien) 91(1–2):12–20
37. Benzer A, Mitterschiffthaler G, Marosi M et al (1991) Prediction of non-survival after trauma: Innsbruck Coma Scale. Lancet 19;338(8773):977–8
38. Born JD, Albert A, Hans P, Bonnal J (1985) Relative prognostic value of best motor response and brain stem reflexes in patients with severe head injury. Neurosurgery 16(5):595–601
39. Rappaport M, Dougherty AM, Kelting DL (1992) Evaluation of coma and vegetative states. Arch Phys Med Rehabil 73(7):628–34
40. Stanczak DE, White JG, Gouview WD, et al (1984) Assessment of level of consciousness following severe neurological insult. A comparison of the psychometric qualities of the Glasgow Coma Scale and the Comprehensive Level of Consciousness Scale. J Neurosurg 60(5):955–60

Chapter 4
Clinical Evaluation of Residual Brain Function and Responsiveness in Disorders of Consciousness

Heinrich Binder

Abstract The neuronal processes that sustain consciousness, arousal and awareness interfered within disorders of consciousness remain poorly defined. As a consequence, subjects with DoC are classified (and the residual brain functions are assessed) based on vital and neurological signs and behavioural patterns. Efficient and reliable clinical scales have been designed and are commonly in use for the bedside evaluation of subjects with DoC. Problems in defining consciousness and its behavioural descriptors by unambiguous, common and applicable terms nevertheless still limit the classification of DoC as distinct diagnostic entities.

Keywords Disorder of consciousness • Residual brain functions • Responsiveness • Clinical scales

4.1 Introduction

Disorders of consciousness (DOC) are always the consequence of a severe disturbance of brain function indicating the urgent need for a comprehensive causative analysis of these conditions. On the one hand, there is whole range of different DOCs delicate to delimit and on the other hand a wealth of possible partial characteristics that need to be considered. Basically, DOCs are mental disorders primarily based on, or a consequence of, an organic disease. They are listed in the international statistical classification of diseases and related health problems (ICD) in chapter XVIII/R40.0-.2 (symptoms, signs and abnormal clinical and laboratory findings) but also in chapter V/F0.5, F06.5 (mental and behavioural disorders) and only marginal in chapter VI (diseases of the nervous system). This indicates that DOCs as of yet have not been appreciated as a distinct coherent diagnostic entity.

H. Binder
Department für Klinische Medizin und Präventionsmedizin, Neurological Center,
Karl Landsteiner Institut für Neurorehabilitation und Raumfahrtneurologie, Vienna, Austria

Neurological Centre, Otto Wagner Hospital, Vienna, Austria
e-mail: heinrich.dr.binder@outlook.com; heinrich.binder@wienkav.at

M.M. Monti, W.G. Sannita (eds.), *Brain Function and Responsiveness in Disorders of Consciousness*, DOI 10.1007/978-3-319-21425-2_4

4.2 The Problem of Defining Consciousness

There are various approaches to the term consciousness. There are many philosophical and also deep analytical attempts to explain, but these are impractical for daily clinical use. Considering that a 'concept' is defined as 'to mean something explicitly', the concept of consciousness needs refinement in terms of advancing determination. Particularly in medical terms, it is necessary to define the concept of consciousness with precise, unambiguous and workable traits in order to have a chance for demarcation of unhealthy/pathological from healthy/physiological condition.

A more oversimplified definition says: consciousness is the state of full awareness of the self and one's relationship to the environment, and it has two components: content and arousal [1]. To ascertain the presence of consciousness in a patient beyond any doubt, it is necessary to obtain such information directly from them. In doing so, two points must be kept in mind: under certain circumstances, the patient is not able to give such information because of inability to verbalise or otherwise express themselves in case of language disorder or paralysis. In addition, it may also be that some patients might be unwilling to cooperate. It is the skill and knack of the examiner to ask the right questions, to observe behaviour, to try to understand and to interpret. The whole trick is to ensure objectivity and to get into another's mind concurrently. Therefore, Thomas Nagel legitimately asks 'What is like to be a bat?' [2] What is the issue here? It is about the content including cognitive and affective functions based on the sum of all functions at the cortical level. In this regard, it is interesting to recall Damasio's hypothesis of consciousness. He hypothesised the unconscious perception of one's own body as the bottom line, the proto-self and the sustained perception of the environment as the core self. He proceeded on the assumption of different neuronal mappings interacting with each other: object mapping, mapping of the organism – the Self– and mapping of the relation between organism and object. Furthermore, he postulated the autobiographic self which means language, long-term memory and thinking [3]. Considering this and other positions not mentioned here for brevity, it can be said that consciousness is a mongrel label employed to refer to different concepts and denoting different phenomena [4]. Just because consciousness, as a global phenomenon, results from a complex interaction of many processes not directly relatable to specific neuroanatomical structures, it does not mean that single aspects of consciousness might not be localizable. Therefore, a kind of fractional loss of consciousness might occur if a lesion of any kind whatsoever disrupts one of the networks at strategically important point [1]. Blind sight, neglect and anosognosia should be noted as examples.

Arousal, the degree of general cerebral activation or, colloquially, the level of consciousness, is the second component of consciousness next to content. Contrary to content, which is based on the above-mentioned more or less mongrel concept, arousal is bound to specific sets of brainstem and diencephalic pathways normally linked behaviourally to the appearance of wakefulness and sleep. Degree of general cerebral activation is reflected by wakefulness or vigilance which means ability of permanent alertness to oneself and environment. Therein included is attention, the ability for selective concentration on discrete aspects of information. This short discussion is already indicative of easier comprehensibility and conceivability of the arousal aspect of consciousness.

4.3 Disorders of Consciousness: What Is Meant by This?

Disorders of consciousness can relate to the content, as well as to the level, of consciousness. Impairment of the former doesn't imply an impairment of the latter. Conversely, an impaired arousal far exceeding the normal range does affect the content of consciousness. This pertains both hyperarousal and hypoarousal states. Disorders of consciousness commonly relate almost solely to hypoarousal conditions. Therefore, the altered arousal state is part of a more or less varied complex syndromatic collection of cognitive, affective, sensory, motoric and autonomic impairments. To consider this is important since structures responsible for consciousness differ from those regulating the last mentioned functional areas.

Basically, alterations of consciousness can appear acutely or chronically. Clouded consciousness, delirium, obtundation, stupor, coma and locked-in syndrome are mostly acute conditions. Conversely, hypersomnia, abulia, minimally conscious state and vegetative state count as chronic alterations of consciousness [1].

Acute alterations of consciousness:

- Clouded consciousness is applied to a state of reduced wakefulness, inattention, usually incomplete orientation to time and place and drowsiness alternating with agitation and hyperexcitability.
- Delirium is defined in DSM-IV [5] as a disturbance of consciousness with reduced ability to focus, sustain, or shift attention; a change in cognition; or the development of a perceptual disturbance that is not better accounted for by a preexisting, established or evolving dementia. Delirium develops usually over hours to days, tends to fluctuate during the course of the day and rarely last more than 1 week in full manifestation. Three subtypes are to be distinguished: hyperactive, hypoactive and mixed delirium [6, 7].
- Obtundation is a state of severe drowsiness. Patients tend to sleep most of the day. They can be aroused for short time by moderate stimulation with drowsiness, lessened interest in the environment and slowed responses to stimulation during these phases.
- Stupor is a condition of deep sleep or similar behaviour. Patients arouse only after strong and repeated stimulation immediately relapsing to the former unresponsive state.
- Coma is a commonly acute state. Patients are unarousable despite vigorous stimulation. Nevertheless, different motor responses to painful stimulation can occur, albeit untargeted, not at least also depending on the underlying disease or lesion. Patients that survive coma can evolve to a chronic condition such as the vegetative state or the minimally conscious state.

- Locked-in syndrome is a condition of quadriplegia and paralysis of lower cranial nerves: patients can communicate only by blinking or vertical eye movements. Typically, it occurs after brainstem infarct or haemorrhage. The symptomatology can become chronic or gradually ameliorate over a longer period of months to years [8].

Chronic alterations of consciousness (see Table 4.1):

- Hypersomnia is characterised by excessive sleep. The patient awakens, when stimulated, for short periods of time with dulled consciousness, but otherwise presents normal cognitive functions. A hypothalamic dysfunction comes aetiologically into question [9].
- Abulia is an apathetic state in which patients show slowed reactions and no initiative neither for communication nor motor activities.
- Akinetic mutism is a condition of silent, alert-appearing immobility with normal sleep–wake cycle but without recognisable mental activity and spontaneous motor activity. Abulia, as well as akinetic mutism, usually occurs after bilateral frontal lesions.
- Minimally conscious state (MCS) is a condition where patients show limited but clearly discernible evidence of self-awareness or environmental awareness, as demonstrated in a sustained or reproducible manner. Examples include eye tracking and sustained fixation in direct response to salient stimuli. Also appropriate smiling or crying in response to the linguistic or visual content of emotional topics or stimuli is possible as well as vocalisations or gestures in response to the content of questions. Regarding motoric functions, reaching or touching objects is also possible [10].
- Vegetative state and its synonyms apallic syndrome and unresponsive wakefulness [11, 12] need several days to some weeks to develop its phenomenology. Patients classified with this syndrome show no evidence of the self or their environment. They are incapable of interacting with others and show no sustained, reproducible or purposeful voluntary behavioural responses to visual, auditory, tactile or noxious stimuli. As far as clinically estimated, there exists no evidence of language comprehension or expression. There exists an intermittent wakefulness characterised by sleep–wake cycles. However, cranial nerve reflexes and autonomic and hypothalamic functions, sufficient for survival with medical or nursing care, are preserved. Vegetative state as well as minimally conscious state are the consequence of severe structural or metabolic brain injuries. Typical example for the former is the traumatic brain injury with descending or ascending herniation of brainstem structures. The most frequent metabolic cause is the hypoxic encephalopathy in consequence of cardiac irregularity or cardiac arrest [13].

Table 4.1 gives an overview of the most important disorders of consciousness syndromes [10].

Table 4.1 Comparison of clinical features associated with coma, vegetative state, minimally conscious state and locked-in syndrome [10]

Condition	Consciousness	Sleep/wake	Motor function	Auditory function	Visual function	Communication	Emotion
Coma	None	Absent	Reflex and postural responses only	None	None	None	None
Vegetative state	None	Present	Postures or withdraws to noxious stimuli	Startle	Startle	None	None
			Occasional Non-purposeful movement	Brief orienting to sound	Brief visual fixation		Reflexive crying or smiling
Minimally conscious state	Partial	Present	Localises noxious Stimuli	Localises sound location	Sustained visual fixation	Contingent vocalisation	Contingent smiling or crying
			Reaches for objects	Inconsistent command following	Sustained visual pursuit	Inconsistent but intelligible verbalization or gesture	
			Holds or touches objects in a manner that accommodates size and shape				
			Automatic movements (e.g. scratching)				
Locked-in syndrome	Full	Present	Quadriplegic	Preserved	Preserved	Aphonic/anarthric	Preserved
						Vertical eye movement and blinking usually intact	

4.4 How to Assess Patients with Disorders of Consciousness

The evaluation of patients with acute disorders of consciousness is subdivided into three parts. First, it is necessary to determine the level of consciousness. For a preliminary assessment, the Glasgow Coma Scale (GCS) is appropriate in most cases (Table 4.2). This, however, does not release from the obligation to careful detailed psychopathologic examination at a later time. Second, the so-called vital signs must be checked, and appropriate measurements must be done. Third, a careful physical examination and a particularly detailed neurological examination must be conducted concerning presence or absence of focality of disorder, particularly, specific involvement of cortical or brainstem structures on the one hand and the level of dysfunction within the rostrocaudal neuraxis on the other hand, with the latter being particularly concerned with life-threatening brainstem herniation.

Subsequent to the initial assessment of consciousness and vital emergency care, a detailed examination of brain functions should be conducted. It starts out with observation and specific psychopathologic interview. The extent of the latter depends on the patient's communicative ability and relates primarily to noopsychic functions. The interview begins with the assessment of psychomotorics, in other words the overall dynamic impression. After that, the noopsychic functions must be described in detail concerning the level and content of consciousness, orientation, attention, short- and long-term memory, process as well as content of thinking and, not least, perception disorders like hallucinations.

Subsequently, a detailed neurological examination focusing on pupillary and optomotor responses, general motor and autonomic functions is mandatory.

Table 4.2 Glasgow Coma Scale

Eye opening	4. Opens eyes spontaneously
	3. Opens eyes to voice
	2. Opens eyes to pain
	1. No eye opening
Best motor response	6. Obeys commands
	5. Localises to pain
	4. Withdraws to pain
	3. Abnormal flexor response
	2. Abnormal extensor response
	1. No movement
Best verbal response	5. Appropriate and oriented
	4. Confused conversation
	3. Inappropriate words
	2. Incomprehensible sounds
	1. No sounds

GCS > 13: mild brain injury, 9–12: moderate brain injury, <8: severe brain injury

The specific observation of spontaneous posture and motor function, mainly in response to painful stimulation, is key to the assessment of the consciousness disturbance. Spontaneous-sustained fumbling fingers and wiping hand movements are typically observed in cases of delirious states. Gradually decreasing level of consciousness is paralleled by increasingly slowed and undirected movements both spontaneously and in reaction to pain. Some postures are typical of increasing intracerebral pressure and descending brain herniation. It starts with progressing plantar flexion of feet with increasing extensor tone in both lower extremities. Following increasing flexor tone occurs in the upper extremities which results in the so-called decorticate posturing. In chronological sequence, the flexor tone in the upper extremities changes to an extensor tone resulting in so-called decerebrate posture. Concurrently to these changes in tone and posture motion sequences change towards posture enforcing increasing mass – and synergistic spontaneous and pain-induced movements. A different posture occurs in the rare case of ascending herniation with extensor posture of the upper extremities and flaccid paralysis of the legs. It is necessary to be aware of the fact that assessment of cortical functions, and particularly of focal signs, gets lost in parallel to the previously mentioned progress with increasing disinhibition of autonomic brainstem functions. This characteristic course of posture and tone can be seen also as a consequence of metabolic disorders if not necessarily in full manifestation, sometimes only adumbrated, lacking characteristic spastic rigidity with frequently proven focal signs.

The next step is the examination of the pupils and the oculomotor system including the so-called brain stem reflexes to receive information about the integrity of the centres inside the midbrain and the pons. Commonly, the examination starts with the evaluation of the position and tone of the eyelids. Decreasing consciousness and coma are accompanied by reduced muscle tone and lost spontaneous blinking. Two brain stem reflexes involve the eyelids: the blinking reflex in response to bright light and the corneal reflex. Eye position and eye movement come next. Drowsiness, as well impaired consciousness, is associated with slight exophoria in resting position because of failing fixation. However, there are several spontaneous conjugate and disconjugate movements sometimes distinctive of local structural lesions, sometimes unreliable for localization. Examples for spontaneous eye movements caused by local structural lesions are conjugate lateral deviation of the eyes, nonconjugate eye deviation, ocular bobbing, nystagmus and myoclonic jerks. Unreliable movements concerning localization and mostly as a consequence of metabolic or hypoxic injury are roving eye movements, ocular dipping and periodic alternating gaze deviation. Following the observation of spontaneous eye movements, two reflexes should be assessed: the oculocephalic reflex and the caloric vestibuloocular reflex [14]. Both reflexes give information regarding the function of pontine-vestibular connections. The former, also known as doll's eye response, is suppressed by fixation. Therefore, the conjugate eye rotation opposite to passive head rotation will only occur in case of neutralised fixation or comatose patients. In cases of uncertain reaction, sometimes occurring in deep coma, the caloric vestibuloocular response is an alternative. For this, the instillation of cool (30 °C) or warm (44 °C) water inside of the external auditory canal is used. Usually, this leads to brisk ipsiversive (cold) or contraversive (warm) eye deviation, occasionally also nausea and vomiting.

It is important to examine the pupils because size and light reaction are the most important physical signs to distinguish metabolic from structural reasons of disturbed consciousness [1].

Leaving aside the consequences of focal lesion, the size and light response of pupils can give a hint to the level of brain lesions causing impaired consciousness and coma. Small but reactive pupils are observed in case of metabolic encephalopathies but also in diencephalic lesions. Pretectal lesions cause large and fixed pupils. Lesions at the midbrain level lead to midposition of fixed pupils, whilst pontine lesions are associated with the pupil's pinpoint size. The ciliospinal reflex, assessed by pinching the skin of neck or face, can also be used as a spinobulbar response.

The bedside evaluation of patients with subacute or chronic alterations of consciousness consists also of careful neuropsychiatric examination supplemented with special neuropsychologic test attempting to find out behavioural changes indicative of improving consciousness which can be used for further specific treatment measures. Alterations of consciousness appear in the context of extensive cerebral functional disturbances affecting both arousal system and sensory–motor pathways as set out above. To recognise the presence of consciousness poses a great challenge considering severe concomitant motor constraints impeding behaviour necessary for interpersonal exchange. First, stirrings of relapsing of consciousness functions are reactions to so-called visual and auditive autoreferential stimuli. Sustained visual tracking is one of the main clinical signs of responsiveness recovery. Many of the below-mentioned behavioural scales established the capability of visual fixation as a criterion for distinguishing VS from MCS. Its assessment is difficult because of fluctuations of patient responsiveness and different assessment tools. The various scales use different stimuli which range from baby photos (SMART) to different objects (DOCS), persons (WHIM) and light flashes (CNC), or confine themselves to observation of spontaneous eye movements (WNSSP) [15]. Use of mirrors shows a higher frequency of fixation compared to other stimuli and constitutes a very sensitive test [16, 17]. This has been confirmed by eye-tracking recordings [18]. Thereto, similarly autoreferential is the sound localization reflex. The most potent stimulus regarding this is one's own name [19]. The own name is a signal of more significant information content shown in a range of studies. Babies are already automatically responsive to their names demonstrating strong degree of attention [20–22]. Even awaking from general anaesthesia, one's own name creates reactions before pain reactions emerge [23].

There is no doubt that the description of consciousness and its disturbances from a third viewpoint are very difficult and only approximately possible in the form of narrative description. This of course involves the danger of getting lost in details and may lead to difficulties in getting an overview of the disorder. For this reason, it was attempted to define a few determining and measurable characteristics collecting them to scales with different scores.

4.5 The Use of Scales and Scores for Assessment of Disorders of Consciousness

The main problem of most scales is the lack of an operational definition of the construct of consciousness and therefore of its disorders. The American Congress of Rehabilitation Medicine 2010 has given an overview on still very common assessment scales for disorders of consciousness (Table 4.3).

Summarising, the use of the CRS-R is recommended in the first place because of excellent content validity, acceptable standardised administration and scoring procedures. Certain concerns exist with respect to the SMART, WNSSP, SSAM, WHIM and DOCS. These are considered as having good or acceptable content validity, acceptable standardised administration and scoring procedures, but limited evidence with regard to reliability or criterion validity. CNC is estimated with restraint. Finally, RLS85, LOEW, CLOCS, FOUR, INNS and GLS are not recommended (Table 4.4).

4.6 Conclusion

Clinical assessment of cerebral functions in case of disturbed consciousness poses a great challenge because of great behavioural restraints. To overcome these difficulties, careful neuropsychiatric and neuropsychologic examination from a specialist experienced in the field of clinical diagnosis of disturbed consciousness and related syndromes is premise for developing further diagnostic and therapeutic measures. Scores give a rough valuation of consciousness and global prognosis to a certain extent. They cannot replace, however, detailed neuropsychiatric and neuropsychologic assessment with regard to targeted therapeutic measurements.

Table 4.3

Scale		Content	Estimated time	No. scales (items)
CRS-R	Coma Recovery Scale-Revisited [24]	Auditory, visual, motor, oral, communication, arousal	25	8 (23)
SSAM	Sensory Stimulation Assessment Technique [25]	Auditory, vision, tactile, olfactory, gustatory, eye opening, motor vocalisation	30	5 (15)
WNSSP	Western Neurosensory Stimulation Profile [26]	Visual tactile, olfactory, arousal/attention, auditory, expressive communication	45	5 (32)
SMART	Sensory Modality Assessment Technique [27]	Auditory, vision, tactile, olfactory, gustatory, wakefulness, motor, communication	60+	8 (8)
WHIM	Wessex Head Injury Matrix [28]	Basic behaviours, social/ communication, attention/ cognitive, orientation/memory	30–120	4 (58)
DOCS	Disorders of Consciousness Scale [29]	Auditory, visual, tactile, sensory, swallowing, olfactory	45	1 (23)
CNC	Coma/Near-Coma Scale [30]	Visual, auditory, command following, threat response, olfactory, tactile, pain, vocalisation	10	8 (11)
CLOCS	Comprehensive Levels of Consciousness Scale [31]	Eye responses, motor, posture, communication, general responsiveness	5	7 (7)
LOEW	Loewenstein Communication Scale [32]	Mobility, respiration, visual, auditory, communication	25	5 (25)
RLS85	Swedish Reaction Level Scale-1985 [33]	Responsiveness	15	1 (1)
FOUR	Full Outline of Unresponsiveness Score [34]	Eye response, motor response, respiration, brainstem reflexes	10	4 (4)
GLS	Glasgow–Liege Coma Scale [35]	Eye, verbal, motor, brainstem reflexes	5	4 (4)
INNS	Innsbruck Coma Scale [36]	Eye responses, auditory, pain, posture, oral	10	1 (8)

Table 4.4 Subscale diagnostic criteria for VS, MCS and EMCS from the CRS-R [37, 38]

Auditory function scale	Diagnosis	Scale
	MCS	4: Consistent movement to command
	MCS	3: Reproducible movement to command
	VS	2: Localization to sound
	VS	1: Auditory startle
	VS	0: None
Visual function scale	MCS	5: Object recognition
	MCS	4: Object localization: reaching
	MCS	3: Visual pursuit
	MCS	2: Fixation
	VS	1: Visual startle
	VS	0: None
Motor function scale	EMCS	6: Functional object use
	MCS	5: Automatic motor response
	MCS	4: Object manipulation
	MCS	3: Localization to noxious stimulation
	VS	2: Flexion withdrawal
	VS	1: Abnormal posturing
	VS	0: None/flaccid
Oromotor/verbal function scale	MCS	3: Intelligible verbalization
	VS	2: Vocalisation/oral movement
	VS	1: Oral reflexive movement
	VS	0: None
Communication scale	MCS	2: Functional: accurate
	MCS	1: Non-functional: intentional
	VS	0: None
Arousal scale	VS?	3: Attention
	VS?	2: Eye opening without stimulation
	VS?	1: Eye opening with stimulation
	VS	0: Unarousable

Abbreviations: *CRS-R* Coma Recovery Scale-Revisited, *EMCS* emergence from minimal conscious state, *MCS* minimally conscious state, *VS* vegetative state

References

1. Posner JB, Saper CB, Schiff ND, Plum F (2007) Plum and Posner's diagnosis of stupor and coma. Oxford University Press, Oxford/New York
2. Nagel T (1974) What is like to be a bat? Philos Rev 83(4):435–450
3. Dammasio AR (1999) The feeling of what happens: body and emotion in the making of consciousness. Hartcourt Brace, New York
4. Block N (1995) On a Confusion about a Function of Consciousness. The Behavioral and Brain Sciences 18(2):227–287
5. Frances A, Pincus HA, First MB (1994) Diagnostic and statistical manual of mental disorders: DSM IV, 4th edn

6. Liptzin B, Levkoff SE (1992) An empirical study of delirium subtypes. Br J Psychiatry 161:843–845
7. Meagher DJ, O'Hanlon D, O'Mahony E et al (2000) Relationship between symptoms and motoric subtype of delirium. J Neuropsychiatry Clin Neurosci 12:51–56
8. Smith E, Delargy M (2005) Locked-in syndrome. BMJ 330:406–409
9. Billiard M, Dauvilliers Y (2001) Idiopathic hypersomnia. Sleep Med Rev 5:349–358
10. Giacino JT, Ashwal S, Childs N, Cranford R, Jennett B, Katz DI, Kelly JP, Rosenberg JH, Whyte J, Zafonte RD, Zasler ND (2002) The minimally conscious state definition and diagnostic criteria. Neurology 58:349–353
11. Laureys S, Celesia GG, Cohadon F et al (2010) Unresponsive wakefulness syndrome: a new name for the vegetative state or apallic syndrome. BMC Med 8:68
12. Monti MM, Laureys ST, Owen AM (2010) The vegetative state. BMJ 341:c3765
13. Binder H (2014) Neurorehabilitative interventions in the acute stage of diseases. In: Selzer ME, Clarke S, Cohen LG, Kwakkel G, Miller RH (eds) Textbook of neural repair and rehabilitation, 2nd edn. Cambridge University Press, Cambridge, UK
14. Kwan-ChunTing W, PerezVelazquez JLP, Cusimano MD (2014) Eye movement measurement in diagnostic assessment of disorders of consciousness. Front Neurol 5:137 | 2
15. Majerus S, Gill-Thwaites H, Andrews K, Laureys S (2005) Behavioral evaluation of consciousness in severe brain damage. Prog Brain Res 150:397–413
16. Vanhaudenhuyse A, Schnakers C, Bredart S, Laureys S (2008) Assessment of visual pursuit in post-comatose states: use a mirror. J Neurol Neurosurg Psychiatry 79(2):223
17. Di H, Nie Y, Hu X, Tong Y, Heine L, Wannez S, Huang W, Yu D, He M, Thibaut A, Schnakers C, Laureys S (2014) Assessment of visual fixation in vegetative and minimally conscious states. BMC Neurol 14:147
18. Trojano L, Moretta P, Loreto V, Cozzolino A, Santoro L, Estraneo A (2012) Quantitative assessment of visual behavior in disorders of consciousness. J Neurol 259(9):1888–1895
19. Cheng L, Gosseries O, Ying L, Hu X, Yu D, Gao H, He M, Schnakers C, Laureys S, Di H (2013) Assessment of localisation to auditory stimulation in post-comatose states: use the patient's own name. BMC Neurol 13:27
20. Mandel DR, Jusczyk PW, Pisoni DB (1995) Infants' recognition of the sound pattern of their own names. Psychol Sci 6:314–317
21. Kawahara J, Yamada Y (2004) Does one's own name attract visual attention? Vis Cogn 11:997–1017
22. Harris CR, Pashler H (2004) Attention and the processing of emotional words and names: not so special afterall. Psychol Sci 15:171–178
23. Kurtz D, Trapp C, Kieny MT, Wassmer JM, Mugnaioni MD, Pack A et al (1977) Study of recovery and the post-anaesthetic period. Rev Electroencephalogr Neurophysiol Clin 7:62–69
24. Giacino JT, Kalmar K, Whyte J (2004) The JFK Coma recovery Scale-Revisited: measurement characteristics and diagnostic utility. Arch Phys Med Rehabil 85:2020–2029
25. Rader MA, Ellis DW (1994) The Sensory Stimulation Assessment Measure (SSAM): a tool for early evaluation of severely brain- injured patients. Brain Inj 8:309–321
26. Ansell BJ, Keenan JE (1989) The Western neuro sensory Stimulation Profile: a tool for assessing slow- to recover head-injured patients. Arch Phys Med Rehabil 70:104–108
27. Gill-Thwaites HH (1997) The sensory modality assessment rehabilitation technique- a tool for assessment and treatment of patients with severe brain injury in a vegetative state. Brain Inj 11:723–734
28. Shiel A, Horn SA, Wilson BA, Watson MJ, Campbell MJ, McLellan DL (2000) The Wessex head Injury Matrix (WHIM) main scale a preliminary report on a scale to assess and monitor patient recovery after severe head injury. Clin Rehabil 14:408–416
29. Pape TL, Senno RG, Guernon A, Kelly JP (2005) A measure of neurobehavioral functioning after coma. Part II: clinical and scientific implementation. J Rehabil Res Dev 42:19–27
30. Rappaport M, Dougherty AM, Kelting DL (1992) Evaluation of coma and vegetative state. Arch Phys Med Rehabil 73:628–634

31. Stanczak DE, White JG, Gouview WD et al (1984) Assessment of level of consciousness following severe neurological insult: a comparison of the psychometric qualities of the Glasgow Coma Scale and the Comprehensive Level of Consciousness Scale. J Neurosurg 60:955–960
32. Borer-Alafi N, Gil M, Sazbon L, Korn C (2002) Loewenstein communication scale for the minimally responsive patient. Brain Inj 16:593–609
33. Stalhammar D, Starmark JE, Holmgren E et al (1988) Assessment of responsiveness in acute cerebral disorders: a multicentre study on the reaction level scale (RLS 85). Acta Neurochir 90:73–80
34. Wijdicks EF, Bamlet WR, Maramattom BV, Manno EM, McClelland RL (2005) Validation of a new coma scale: the FOUR score. Ann Neurol 58:585–593
35. Born JD, Albert A, Hans P, Bonnal J (1985) Relative prognostic value of best motor response and brain stem reflexes in patients with severe head injury. Neurosurgery 16:595–601
36. Benzer A, Traweger C, Ofner D, Marosi M, Luef G, Schmutzhard E (1995) Statistical modelling in analysis of outcome after trauma Glasgow-Coma-Scale and Innsbruck-Coma-Scale. Anesthesiol Intensivmed Norfallmed Schmerztrher 30:231–234
37. Kalmar K, Giacino JT (2005) The JFK Coma Recovery Scale- revisited. Neuropsychol Rehabil 15:454–460
38. Cruse D, Chennu S, Chatelle C, Fernandez-Espejo D, bekkinschtein TA, Pickard JD, Laureys S, Owen AM (2012) Relationship between etiology and covert cognition in the minimally conscious state. Neurology 78:816–822

Chapter 5
Measuring Consciousness Through Imaging

Carol Di Perri, Jitka Annen, Georgios Antonopoulos, Enrico Amico, Carlo Cavaliere, and Steven Laureys

Abstract At present, the global hallmark of impaired consciousness appears to be a reduced metabolism in a widespread frontoparietal network and a multifaceted dysfunctional connectivity architecture characterized by intra-/internetwork altered connectivity, both in the sense of decrease and increase.

However, despite ongoing efforts, the mechanisms underlying the emergence of consciousness are not thoroughly understood. Moreover, recent studies highlighted difficulty in forming a diagnosis in patients with disorders of consciousness (DOC) based only on behavioral assessments.

We here aim at providing an overview of the insights that functional and structural neuroimaging techniques can provide in comprehending the mechanism underlying consciousness and assessing patients with DOC.

Keywords DOC • Functional neuroimaging • PET • Resting-state fMRI • DTI

5.1 Introduction

Back in the 1950s, improved health care in the intensive care unit allowed patients to sustain heart function and systemic circulation after traumatic and nontraumatic brain injuries. Ever since, some surviving patients were found to suffer from altered states of consciousness that were never encountered before [1]. These were mainly characterized by prolonged states of dissociation of the two components of consciousness: wakefulness and awareness [2]. Wakefulness, or arousal, refers to the level of alertness which is necessary but not sufficient to sustain consciousness [2].

Awareness refers to the ability to live experiences of any kind, comprising subjective experiences such as sensations, thoughts, emotions, memory, imagination, and other major psychological processes [3]. Awareness has been further subdivided into internal awareness (i.e., awareness of self) and external awareness (i.e., awareness of the environment) [4]. Wakefulness can be clinically detected by

C. Di Perri (✉) • J. Annen • G. Antonopoulos • E. Amico • C. Cavaliere • S. Laureys
Coma Science Group, Neurology Department, Cyclotron Research Center,
University of Liège, Liège, Belgium
e-mail: diperric@gmail.com

M.M. Monti, W.G. Sannita (eds.), *Brain Function and Responsiveness in Disorders of Consciousness*, DOI 10.1007/978-3-319-21425-2_5

opened eyes. On the other hand, there is no singular marker of awareness, yet its presence can be clinically deduced from a range of behaviors (e.g., responses to command, visual pursuit) which indicate that an individual can perceive self and surroundings [5].

Coma is a state characterized by the inability to support wakefulness and awareness of the surroundings and of the self, lasting no longer than a few weeks. Patients who awaken (i.e., open their eyes) but remain unresponsive to any external stimulation are considered to be in a vegetative state (VS) [6], a condition now also coined as unresponsive wakefulness syndrome (UWS) [7]. It was only in 2002 that the medical community recognized another entity which characterized patients showing fluctuating yet reproducible behavioral signs of consciousness; these patients were said to be in a minimally conscious state (MCS) [8]. The heterogeneity of MCS has been recently recognized, and it has been proposed to subcategorize this group based on the level of complexity of the observed behavioral responses into MCS minus and MCS plus, of which the latter shows the ability to understand language and therefore to follow simple commands [9, 10]. Patients are considered to emerge from an MCS (EMCS) once reliable communication and object use in a functional manner are restored [8]. Patients in a locked-in syndrome (LIS) are conscious but have lost voluntary motor control, except for small eye movements, making it possible for them to answer yes-no questions. More rarely, LIS patients can lose the control of all their voluntary muscles, including extrinsic eye muscles (complete LIS) [11]. In these patients, the only evidence for preserved consciousness may be their ability to communicate via assisting technologies (i.e., fMRI, EEG, or evoked potentials). As such, the term "functional LIS" has been recently proposed to describe those patients with a dissociation between extreme behavioral motor impairment and the identified preserved higher cognitive abilities only detectable by functional imaging techniques [9].

Despite considerable research, disorders of consciousness (DOC) remain poorly understood in regard to their neural basis, clinical recognition, and long-term outcome [12–14]. Indeed, even if the border zone between patients in a VS/UWS and MCS is at present well delimited, bedside assessment of consciousness is intrinsically gated by behavioral responsiveness. Numerous studies highlight the difficulty in forming a diagnosis for DOC patients when this is established merely on behavioral assessments [15–17]. It is now increasingly recognized that the absence of purposeful behaviors cannot be taken as a definitive proof of the absence of consciousness, that is, the lack of responsiveness does not necessarily imply lack of awareness. Indeed, neurophysiological and functional neuroimaging data indicate that a subset of behaviorally unresponsive patients retain patterns of cerebral processing similar to those observed in conscious patients (functional LIS) [15]. This knowledge has led to the search for nonclinical assessment techniques in order to better understand brain function in these patients and to overcome the limits of behavioral assessment in the detection of possible retained consciousness in unresponsive patients.

We here aim at providing an updated overview of the application and contribution of structural and functional neuroimaging techniques for the understanding of brain functions underlying consciousness and the assessment of patients with DOC. We will first discuss the information we obtain about the brain's metabolic activity (with positron emission tomography-PET) and its hemodinamic functions (with functional magnetic resonance imaging - fMRI). We will focus on resting-state fMRI studies, i.e. in the absence of external or internal stimuli. We will then discuss the information we obtain about the brain's structural properties (with conventional MRI and diffusion-weighted MRI). To our mind, an integration of functional and structural neuroimaging techniques, and more generally of the different technical modalities, is important to gain a holistic vision of the underlying pathophysiology of disorders of consciousness in general and to promote single-patient management in particular.

5.2 Functional Neuroimaging in DOC

Studies using conventional brain structural imaging have shown highly variable and heterogeneous findings in patients with DOC, suggesting that awareness cannot be strictly related to an anatomic region. This induced researchers to adopt alternative neuroimaging techniques in order to investigate the role of brain function in the alterations of consciousness [18].

5.2.1 Positron Emission Tomography

Positron emission tomography (PET) is one of the first functional neuroimaging techniques used to investigate brain function in patients with DOC. It assesses brain activity and function by recording the emission of positrons from radioactively labeled molecules. When the chosen molecule is ^{18}fluorodeoxyglucose (FDG), the concentration of tracer indicates glucose uptake, indirectly representing cerebral functioning, whereas the $H_2{}^{15}O$-radiolabeled PET, reflecting blood flow, is usually used to detect activation during active and passive paradigms [19].

In comatose patients, PET studies show a reduced brain metabolism up to 50–70 % of normal range [1]. After recovery from postanoxic coma, cerebral metabolic rates for glucose showed no drastic recovery, exhibiting glucose rates of only up to 75 % of the normal values [20]. Furthermore, no established relation between cerebral metabolic rates of glucose or oxygen and patient outcome has been found [20]. In addition, a global decrease in cerebral metabolism is not unique for coma, as it can be associated with several different situations with temporary loss of consciousness such as deep sleep [21], general anesthesia [22], and Cotard's syndrome [23].

VS/UWS patients are characterized by a reduced global metabolism as compared with healthy subjects. However, recovery of consciousness did not necessarily coincide with resumption of global metabolic activity [17]. This observation suggested that awareness is not related to global brain metabolism [24], or at least the association between global brain metabolism and awareness is not absolute. Resting-state

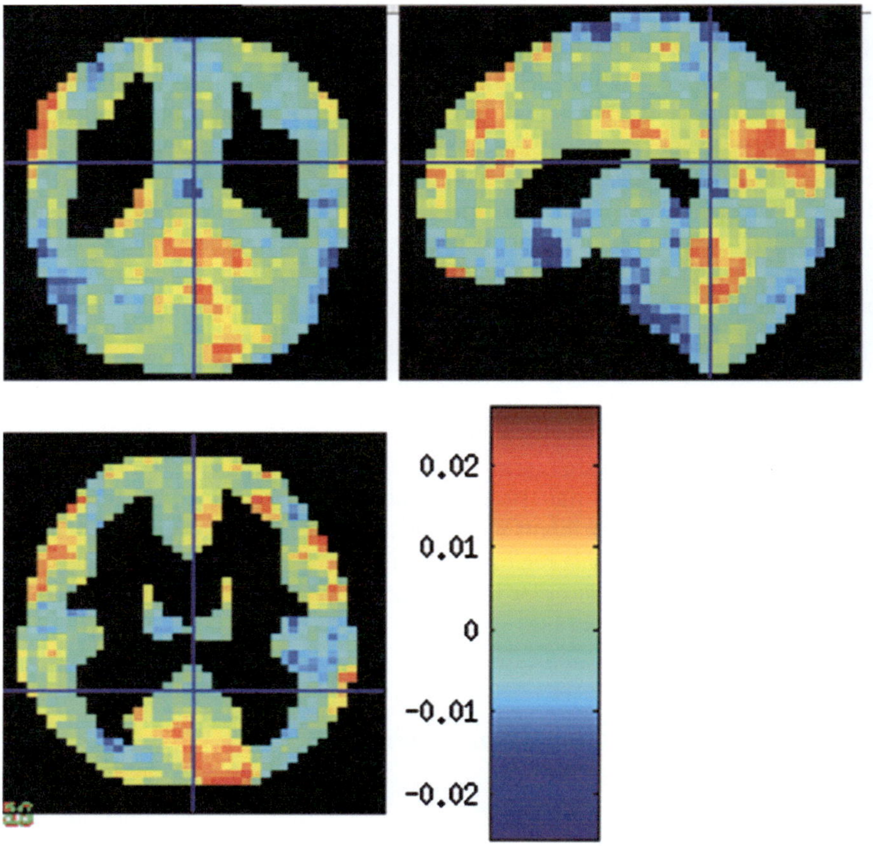

Fig. 5.1 A PET probability map of the hypometabolic areas (*in red*) found in VS/UWS compared to healthy controls

voxel-based PET studies have indeed demonstrated that impairment of awareness was related to decreased glucose uptake in specific brain areas, consisting of a large-scale frontoparietal network encompassing the polymodal associative cortices [25] (Fig. 5.1). In addition, the connections of this network with thalamic nuclei seem also important as suggested by the recovery of thalamocortical activity in a VS/UWS patient who had subsequently recovered consciousness [26]. Moreover, PET studies using passive auditory and noxious stimulation showed a particular disconnection between the primary sensory areas and these higher-order associative cortices [27], which are thought to be required for conscious perception [2]. In simple words, in VS/UWS, any stimulus (auditory, tactile, sensory, etc.) can reach the primary sensory cortex; however, it cannot be processed by the associative cortices which have been proposed to be necessary for conscious perception. As expected regarding the clinical condition of MCS patients, they show a partial metabolic preservation of this large-scale associative frontoparietal network [28]. For example, passive noxious stimuli have elicited the activation of associative areas related to pain processing in

MCS patients similarly as in normal controls, suggesting a potential pain perception capacity in this patient category [19].

The abovementioned studies have suggested the importance of the associative cortices, rather than the primary sensory ones, in the emergence of consciousness. The widespread frontoparietal network encompassing associative cortices, commonly impaired during altered states of consciousness, has been recently functionally subdivided into two different networks: the extrinsic awareness network and intrinsic awareness network [29]. The extrinsic network, or executive control network (ECN), encompasses the lateral frontoparietal regions and is active during attention-demanding task. The intrinsic network, also called default mode network (DMN), encompasses the midline frontoparietal regions (anterior cingulate/mesial prefrontal cortex and posterior cingulate/precuneus) and is active during rest [30]. More specifically, several event-related fMRI studies have demonstrated the ECN and the DMN are linked to external awareness and internal awareness, respectively. For example, by giving somatosensory stimuli to subjects, it has been shown that at a certain intensity, one could perceive the stimulus only when the external awareness network (ECN) was activated. Conversely, a stimulus at the same intensity could not be consciously perceived when the internal awareness network (DMN) was activated [31]. Furthermore, a number of autoreferential stimuli such as the own name and own face pointed to the critical role of the midline structures (DMN) for self or internal awareness [32–35].

The key role of these two networks in the emergence of awareness has been further supported by a recent study on DOC patients in comparison with healthy controls showing metabolic impairment of the DMN and ECN in VS/UWS patients, preserved metabolism of the ECN in MCS patients, and a partial metabolic recovery of both the DMN and ECN only in EMCS [36]. These latter findings suggested that the (partial) metabolic recovery of both networks might be a feature of the emergence from MCS.

In terms of diagnostic accuracy, cerebral metabolism has shown to be able to specifically and reliable differentiate VS/UWS from LIS patients and healthy controls [37]. PET investigation of brain function in MCS has been in accordance with the clinical subcategorization into MCS plus and MCS minus. As expected, patients in MCS plus exhibited preserved metabolism in language comprehension-related areas compared with patients in MCS minus [10]. Furthermore, a recent clinical investigation of the diagnostic accuracy of multimodal neuroimaging methods in disorders of consciousness suggested that cerebral PET could be used to complement bedside examinations and predict long-term recovery of patients in VS/UWS. Indeed, PET was employed to correctly predict outcome in 75 of 102 patients, while active-task fMRI predicted the outcome in 36 of 65 patients [38].

5.2.2 *Functional Magnetic Resonance Imaging (fMRI)*

fMRI relies on the natural paramagnetic properties of deoxygenated hemoglobin to infer changes of blood oxygenation in the brain associated with neural activity (blood oxygenated-level dependent (BOLD) signal). As fMRI does not require the use of X-rays or radiotracer injection, and for its superior spatial and temporal resolution in comparison with PET, it is being increasingly used for detecting brain activity. In the last decades, active-task fMRI has been largely used in patients with DOC to detect brain activity related to residual cognition and awareness. In some cases, it was even employed to establish a two-way communication without requiring any external output, a so-called brain–computer interface [17]. However, active-task fMRI paradigms can suffer from false-negative results because absent command-related brain activation does not allow to infer that awareness is not present [39]. To overcome these limits, resting-state fMRI has been used increasingly.

Resting-state fMRI is particularly suitable for patients with DOC or generally in noncompliant patients as it surpasses the need for patients' participation (the subject is only asked to be still in the scanner without thinking of anything), and it does not require sophisticated experimental setup to administer external stimuli. Resting-state fMRI is used to investigate the spontaneous temporal coherence in BOLD fluctuations related to the amount of synchronized neural activity (i.e., functional connectivity) existing between different brain locations [40]. Brain regions showing the same time course are believed to be functionally connected and form the so-called resting-state functional networks.

Several spatially consistent networks have been identified, such as the auditory network, the visual network, the salience network, and the cerebellum network [41]. However, the DMN has been investigated the most intensively by the scientific community. Given that the DMN seems to be linked to internally oriented cognitive content, such as mind wandering and autobiographical memory recall [4, 42, 43], it is not surprising that it has raised interest in the field of consciousness.

Recent resting-state fMRI studies have shown a breakdown of functional connectivity within the DMN decreasing with the degree of severity of DOC [44]. The reduced connectivity affected mostly the precuneus, a main node of the DMN considered to be a critical hub for consciousness [1]. Similarly, decreases of DMN connectivity were observed in healthy controls during different states of altered consciousness, like sedation, general anesthesia [45], and deep sleep [46].

Recently, more resting-state networks have been investigated in DOC, such as the bilateral executive control network (ECN), the salience, the sensorimotor, the auditory, the visual, and the cerebellar network. It was found that besides the DMN, also the bilateral external awareness network (ECN) and the auditory system were significantly less identifiable (in terms of spatial and neural properties) in DOC patients compared to healthy controls showing consciousness-level-dependent decreases in functional connectivity across the spectrum of DOC (Fig. 5.2). Furthermore, machine learning classification was used to identify if these networks showed neuronal activity or not, employed to separate healthy controls from patients in DOC with 85 % accuracy [47].

Along with the reduced intranetwork DMN connectivity reported in patients with DOC, an increased connectivity in structures of the limbic system was recently reported that was stronger in VS/UWS than in MCS patients [48]. This hyperconnectivity has been thought to reflect the persistent engagement of residual neural activity in self-reinforcing loops which may impair the normal connectivity pattern

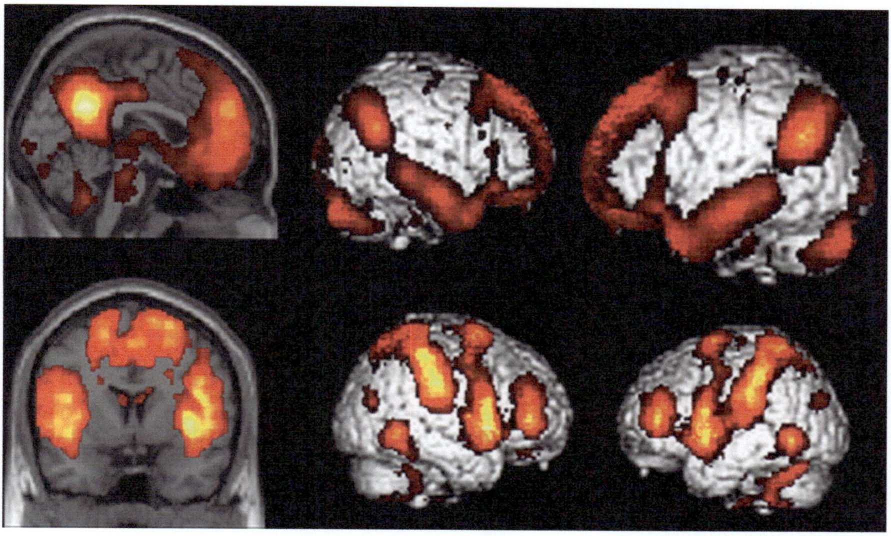

Fig. 5.2 Two anticorrelated resting-state functional networks: default mode network (*top line*) and executive control network (*bottom line*). Statistical maps (*in red*) are superimposed on biplanar and triplanar anatomical slices

[48]. Taking this into account, we have reasons to believe that hypoconnectivity might be only one aspect of a possibly multifaceted and more complex dysfunctional architecture of brain connectivity in altered states of consciousness, not being necessarily the global hallmark of these conditions.

Although these findings show that part of the resting-state brain activity may be related to mental activity and cognition, they appear in contrast with other studies showing the persistence of DMN functional connectivity even during loss of consciousness due to general anesthesia and in VS/UWS patients [49–51]. Although it is known that a small percentage of patients behaviorally in VS/UWS show a residual cognitive activity, this is very unlikely to happen during anesthesia-induced unconsciousness if DMN functional connectivity was solely related to consciousness. These findings suggested that resting-state brain activity might actually have a twofold meaning. An aspect of this activity reflects a fundamental intrinsic property of the brain's anatomical organization that seems to transcend the levels of consciousness, i.e., it is generated by the circulation of spontaneous neural activity along fixed anatomical architecture. The other one seems to be related to cognitive processes that actually characterize the conscious state [52].

A recent study on three monkeys during different levels of consciousness attempted to disentangle these two components and identify the resting-state brain pattern strictly related to consciousness. During the unconscious state, they found resting-state brain activity characterized by few functional connectivity patterns similar to the brain structure and the lack of anticorrelation patterns. Conversely, wakefulness was characterized by the sequential exploration of a richer repertoire of functional connectivity patterns, often dissimilar to anatomical structure, and the presence of anticorrelation between DMN and ECN, suggesting the importance of anticorrelation for the emergence of consciousness. Indeed, since the early studies of resting state, the DMN and the ECN have been shown to be functionally anticorrelated, meaning that when regions belonging to one network show above baseline activity, the regions belonging to the other network are active below their baseline and vice versa [30, 53]. In other words, there is a dynamic interaction between the two functional networks (Fig. 5.2). Such anticorrelation patterns have been observed as well in active-task fMRI studies on motor performance and perceptual discrimination [31], as mentioned above.

More recently, the characteristics of this resting-state activity have been investigated more thoroughly, and it was found that DMN and ECN naturally switch with a mean frequency of 0.05 Hz (range 0.01–0.1 Hz) [4]. Interestingly, this frequency is similar to that of behavioral correlates of external and internal awareness [4]. Activation of DMN was linearly linked to the intensity of internal awareness, whereas activation of ECN was linearly associated with intensity of external awareness [4], suggesting that anticorrelated patterns are functionally significant for mentation and cognition.

However, it should be mentioned the fMRI anticorrelation has been a subject of important debate in literature. It has been argued, for example, that the anticorrelated patterns could arise from the preprocessing procedure when the brain's global activity [54] is regressed out. Nevertheless, as mentioned above, evidence for a physiological origin of the observed anticorrelation is increasing [30, 52, 55], as reduced anticorrelation of these two networks is observed in unconscious states such as anesthesia [45], deep sleep [56], and VS/UWS [50].

Taken together, all the above findings suggest a complex and multifaceted alteration of functional connectivity impaired consciousness, characterized by adapted intra-/internetwork connectivity (both in the sense of decrease and increase).

Unfortunately, fMRI is not suitable to every patient. Its use is virtually impossible for those patients carrying noncompatible MRI devices (pacemakers, metallic implants, etc.), often present in an intensive care setting. Apart from the known contraindication for MRI, the clinical applicability of resting-state fMRI is further limited by several practical issues. For example, patients that show prescan motion activity might need to be anesthetized during the scan. In this case, anesthesia will affect the acquired data which needs to be accounted for during data processing. Another concern is related to the severe brain deformation often observed in these statistical injured patients, which can influence spatial normalization, an essential step for group comparisons. The latter has been partially tackled in previous studies, where a midtemplate [48] or study template [37] was generated by taking into account the mean size of the group, in order to use the same level of deformation in both patient and control groups. Further issues are represented by metal and/or blood artifacts, which are commonly found in severely brain-injured patients and can cause a dropout of signal in the MRI images, and motion artifacts. Moreover, the heterogeneity of the methodological approaches and the lack of a gold standard for resting-state fMRI analyses further hamper the interpretation of these data.

Future work should focus on unraveling the relationship (correlation and anticorrelation) between the resting-state networks in different levels of consciousness, going beyond static functional connectivity measures to the assessment of temporal dynamics of connectivity, namely, looking at the changes of correlation and anticorrelation among networks across time. Finally, the biggest challenge will be the applicability of resting-state fMRI on the single-subject level, as required by clinical practice would require.

5.3 Structural Neuroimaging

5.3.1 Conventional Structural MRI

MRI with conventional sequences (T1-TSE, T2-TSE, FLAIR) is the method of choice to detect brain edema, contusion, hematomas, herniation, hemorrhage, hydrocephalus, or hemorrhagic shearing lesion due to diffuse axonal injuries that are commonly observed in posttraumatic patients (T2* sequences) [18].

Some studies have shown the predictive value of the conventional sequences. For example, the number of lesion detected by FLAIR and T2* sequences has been shown to be inversely correlated with the Glasgow Coma Scale of traumatic patients in a coma. The presence of lesions in the corpus callosum and dorsal midbrain has shown to correlate with reduced chance of recovery in coma patients [57, 58]. However, despite encouraging results, these methods have failed to explain the discrepancy between the structural lesions and the clinical condition observed in some patients, highlighting the lack of specificity and sensitivity of conventional MRI in DOC, which so far has not proven to be a reliable diagnostic tool for this patient category [18].

5.4 Diffusion-Weighted Imaging

Diffusion-weighted imaging (DWI) is an MRI technique used to make inferences about white matter structural integrity, otherwise not detectable by conventional structural imaging. The principle behind DWI is molecular diffusion. Molecules in a fluid diffuse randomly, or isotropic, by means of thermal energy (Brownian motion). However, in the brain, there are physical boundaries, especially along white matter tracts, causing higher diffusion along the axons than perpendicular to them. This is called anisotropic diffusion and allows for measurement of the direction of diffusion and, hence, can be used to make inferences about structural integrity. DWI are often employed to compute fractional anisotropy (FA) and mean diffusivity (MD) and to model fibers/axons (DTI) [59, 60] (Fig. 5.3).

DWI is a powerful technique for in vivo detection of diffuse axonal injury after brain trauma [61] as it is not influenced by sedatives or hypnotics, which do influence clinical exams or fMRI studies. Regarding DOC patients, DTI has been largely

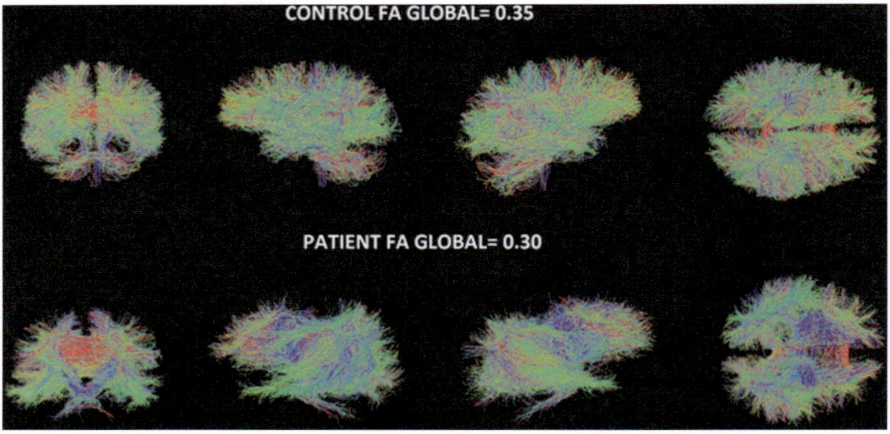

Fig. 5.3 Neuronal long-distance fiber pathways of a patient in comparison with a healthy control. There is evidence of important structural damage; value of whole brain fractional anisotropy average for this patient is 0.30. The level of structural damage quantified by diffusion tensor imaging is consistent with the diagnosis of VS/UWS. Colors indicate directionality of water diffusion: *red* left-right, *green* anterior-posterior, *blue* superior-inferior

applied in traumatic brain injury (TBI) patients. A first study on DTI in diffuse axonal injury was published in 2002 [62] where DTI was performed in five patients with mild traumatic brain injuries within 24 h of injury and ten control subjects. FA was reduced, mainly the internal capsule and corpus callosum, even if they were macroscopically intact. These findings are supported by other more recent studies reporting a significant reduction of FA in both the callosal genu and splenium regions, accompanied by a reduction of the number of splenium-reconstructed fibers in severe TBI [63]. Moreover, the alterations in the genu and splenium of the corpus callosum correlated significantly with the Revised Glasgow Outcome Scale and outcome of these patients. Other studies on severe TBI patients [64–67] showed altered diffusion parameters, mainly FA, in many regions (included anterior and posterior corona radiata, corticospinal tracts, cingulum fiber bundles, external capsule, corpus callosum, etc.). Again, these alterations were negatively correlated with all cognitive domains [65] and clinical outcome [66].

Recent multicentric studies have shown the predictive value of DTI regarding 1-year outcome as it performed better than structural and clinical assessments for both traumatic and anoxic patients [67, 68]. Furthermore, the combination of DTI and MRI spectroscopy (a noninvasive method that is able to measure resonances of different metabolic compounds in the brain) has shown to predict long-term outcome of traumatic patients [69], with a prediction of nonrecovery after 1 year up to 86 % sensitivity and 97 % specificity.

Few studies have tried to characterize microstructural white matter alterations in anoxic patients. Recently, a case report on 76-year-old man with VS/UWS examined with DTI at days 41, 75, 173, and 284 after cardiac arrest reported an early FA reduction in 11 white matter regions that preceded macroscopic MRI and postmortem neuropathological findings [70]. A study performed on fourty-nine cardiac arrest patients [71] demonstrated a significant reduction of axial and radial diffusivity in 19 cerebral white matter regions of interest. Other studies [68, 72], using the same regions of interest (ROI) approach, investigated diffusion alterations in fifty-seven and fourth-nine coma survivors of cardiac arrest, both finding highly dynamic FA values following injury. This alteration predicts unfavorable outcome with a 94 % sensitivity and 100 % specificity [68].

Regarding diagnostic applicability in DOC patients, a recent study used DTI to identify measures that could potentially distinguish VS/UWS and MCS patients [73]. Diffusion in MCS and VS/UWS patients appeared to differ significantly in subcortical white matter and thalamic regions but appeared not to differ in the brainstem. DTI results predicted scores on the Glasgow Coma Scale (GCS) and successfully classified the patients into their appropriate diagnostic categories with an accuracy of 95 % [73]. Furthermore, DWI could characterize etiologic differences in patients in VS/UWS, with abnormalities in the brainstem confined to the traumatic brain-injured group [74].

All the abovementioned studies suggest the relevance of DWI-MRI techniques to quantify white matter integrity and employ the possible application of it for outcome prediction in this patient population.

5.5 Conclusions

In the last decade, the need to objectively measure brain damage associated with DOC and to better understand the mechanism underlying impaired consciousness has boosted the development of nonconventional neuroimaging techniques. Here we have reviewed the basic principles of how the main neuroimaging techniques (PET, fMRI, structural MRI-DTI) provide us with important insights about brain function in DOC patients.

We encourage a multimodal imaging approach to gain a holistic vision of the underlying pathophysiology of DOC in general and to promote single-patient medical management in particular. This will possibly lead to a deeper understanding of the mechanisms underlying consciousness and to a significant drop of misdiagnosis of DOC patients. Finally, among the next challenges is the application of these methodologies at the single subject level, as clinical reality requires.

References

1. Laureys S, Owen AM, Schiff ND (2004) Brain function in coma, vegetative state, and related disorders. Lancet Neurol 3(9):537–546, PubMed Epub 2004/08/25. eng
2. Laureys S (2005) The neural correlate of (un)awareness: lessons from the vegetative state. Trends Cogn Sci 9(12):556–559, PubMed Epub 2005/11/08. eng
3. Cavanna AE, Ali F (2011) Epilepsy: the quintessential pathology of consciousness. Behav Neurol 24(1):3–10, PubMed Epub 2011/03/31. eng

4. Vanhaudenhuyse A, Demertzi A, Schabus M, Noirhomme Q, Bredart S, Boly M et al (2011) Two distinct neuronal networks mediate the awareness of environment and of self. J Cogn Neurosci 23(3):570–578, PubMed Epub 2010/06/03. eng
5. Boly M, Massimini M, Garrido MI, Gosseries O, Noirhomme Q, Laureys S et al (2012) Brain connectivity in disorders of consciousness. Brain Connect 2(1):1–10, PubMed Epub 2012/04/20. eng
6. Plum F, Posner JB (1972) The diagnosis of stupor and coma. Contemp Neurol Ser 10:1–286, PubMed Epub 1972/01/01. eng
7. Laureys S, Celesia GG, Cohadon F, Lavrijsen J, Leon-Carrion J, Sannita WG et al (2010) Unresponsive wakefulness syndrome: a new name for the vegetative state or apallic syndrome. BMC Med 8:68, PubMed Pubmed Central PMCID: PMC2987895. Epub 2010/11/03. eng
8. Giacino JT, Ashwal S, Childs N, Cranford R, Jennett B, Katz DI et al (2002) The minimally conscious state: definition and diagnostic criteria. Neurology 58(3):349–353, PubMed Epub 2002/02/13. eng
9. Bruno MA, Vanhaudenhuyse A, Thibaut A, Moonen G, Laureys S (2011) From unresponsive wakefulness to minimally conscious PLUS and functional locked-in syndromes: recent advances in our understanding of disorders of consciousness. J Neurol 258(7):1373–1384, PubMed Epub 2011/06/16. eng
10. Bruno MA, Majerus S, Boly M, Vanhaudenhuyse A, Schnakers C, Gosseries O et al (2012) Functional neuroanatomy underlying the clinical subcategorization of minimally conscious state patients. J Neurol 259(6):1087–1098, PubMed Epub 2011/11/15. eng
11. Patterson JR, Grabois M (1986) Locked-in syndrome: a review of 139 cases. Stroke 17(4): 758–764, PubMed Epub 1986/07/01. eng
12. Giacino JT, Fins JJ, Laureys S, Schiff ND (2014) Disorders of consciousness after acquired brain injury: the state of the science. Nat Rev Neurol 10(2):99–114, PubMed Epub 2014/01/29. eng
13. Laureys S, Boly M, Maquet P (2006) Tracking the recovery of consciousness from coma. J Clin Invest 116(7):1823–1825, PubMed Pubmed Central PMCID: PMC1483158. Epub 2006/07/11. eng
14. Laureys S (2005) Science and society: death, unconsciousness and the brain. Nat Rev Neurosci 6(11):899–909, PubMed Epub 2005/11/02. eng
15. Schnakers C, Vanhaudenhuyse A, Giacino J, Ventura M, Boly M, Majerus S et al (2009) Diagnostic accuracy of the vegetative and minimally conscious state: clinical consensus versus standardized neurobehavioral assessment. BMC Neurol 9:35, PubMed Pubmed Central PMCID: PMC2718857. Epub 2009/07/23
16. Owen AM, Coleman MR, Boly M, Davis MH, Laureys S, Pickard JD (2007) Using functional magnetic resonance imaging to detect covert awareness in the vegetative state. Arch Neurol 64(8):1098–1102, PubMed Epub 2007/08/19. eng
17. Monti MM, Vanhaudenhuyse A, Coleman MR, Boly M, Pickard JD, Tshibanda L et al (2010) Willful modulation of brain activity in disorders of consciousness. N Engl J Med 362(7): 579–589, PubMed Epub 2010/02/05. eng
18. Tshibanda L, Vanhaudenhuyse A, Boly M, Soddu A, Bruno MA, Moonen G et al (2010) Neuroimaging after coma. Neuroradiology 52(1):15–24, PubMed Epub 2009/10/29. eng
19. Boly M, Faymonville ME, Schnakers C, Peigneux P, Lambermont B, Phillips C et al (2008) Perception of pain in the minimally conscious state with PET activation: an observational study. Lancet Neurol 7(11):1013–1020, PubMed Epub 2008/10/07. eng
20. DeVolder AG, Goffinet AM, Bol A, Michel C, de Barsy T, Laterre C (1990) Brain glucose metabolism in postanoxic syndrome. Positron emission tomographic study. Arch Neurol 47(2):197–204, PubMed Epub 1990/02/01. eng
21. Maquet P, Degueldre C, Delfiore G, Aerts J, Peters JM, Luxen A et al (1997) Functional neuroanatomy of human slow wave sleep. J Neurosci 17(8):2807–2812, PubMed Epub 1997/04/15. eng
22. Alkire MT, Haier RJ, Barker SJ, Shah NK, Wu JC, Kao YJ (1995) Cerebral metabolism during propofol anesthesia in humans studied with positron emission tomography. Anesthesiology 82(2):393–403; discussion 27A. PubMed Epub 1995/02/01. eng

23. Charland-Verville V, Bruno MA, Bahri MA, Demertzi A, Desseilles M, Chatelle C et al (2013) Brain dead yet mind alive: a positron emission tomography case study of brain metabolism in Cotard's syndrome. Cortex 49(7):1997–1999, PubMed Epub 2013/05/15. eng
24. Laureys S, Lemaire C, Maquet P, Phillips C, Franck G (1999) Cerebral metabolism during vegetative state and after recovery to consciousness. J Neurol Neurosurg Psychiatry 67(1):121, PubMed Pubmed Central PMCID: PMC1736451. Epub 1999/08/24. eng
25. Laureys S, Goldman S, Phillips C, Van Bogaert P, Aerts J, Luxen A et al (1999) Impaired effective cortical connectivity in vegetative state: preliminary investigation using PET. Neuroimage 9(4):377–382, PubMed Epub 1999/04/07. eng
26. Laureys S, Faymonville ME, Luxen A, Lamy M, Franck G, Maquet P (2000) Restoration of thalamocortical connectivity after recovery from persistent vegetative state. Lancet 355(9217):1790–1791, PubMed Epub 2000/06/01. eng
27. Laureys S, Faymonville ME, Peigneux P, Damas P, Lambermont B, Del Fiore G et al (2002) Cortical processing of noxious somatosensory stimuli in the persistent vegetative state. Neuroimage 17(2):732–741, PubMed Epub 2002/10/16. eng
28. Laureys S, Perrin F, Faymonville ME, Schnakers C, Boly M, Bartsch V et al (2004) Cerebral processing in the minimally conscious state. Neurology 63(5):916–918, PubMed Epub 2004/09/15. eng
29. Demertzi A, Soddu A, Laureys S (2013) Consciousness supporting networks. Curr Opin Neurobiol 23(2):239–244, PubMed Epub 2013/01/01. eng
30. Fox MD, Snyder AZ, Vincent JL, Corbetta M, Van Essen DC, Raichle ME (2005) The human brain is intrinsically organized into dynamic, anticorrelated functional networks. Proc Natl Acad Sci U S A 102(27):9673–9678, PubMed Pubmed Central PMCID: PMC1157105. Epub 2005/06/25. eng
31. Boly M, Balteau E, Schnakers C, Degueldre C, Moonen G, Luxen A et al (2007) Baseline brain activity fluctuations predict somatosensory perception in humans. Proc Natl Acad Sci U S A 104(29):12187–12192, PubMed Pubmed Central PMCID: PMC1924544. Epub 2007/07/10. eng
32. Perrin F, Schnakers C, Schabus M, Degueldre C, Goldman S, Bredart S et al (2006) Brain response to one's own name in vegetative state, minimally conscious state, and locked-in syndrome. Arch Neurol 63(4):562–569, PubMed Epub 2006/04/12. eng
33. Platek SM, Loughead JW, Gur RC, Busch S, Ruparel K, Phend N et al (2006) Neural substrates for functionally discriminating self-face from personally familiar faces. Hum Brain Mapp 27(2):91–98, PubMed Epub 2005/07/22. eng
34. Staffen W, Kronbichler M, Aichhorn M, Mair A, Ladurner G (2006) Selective brain activity in response to one's own name in the persistent vegetative state. J Neurol Neurosurg Psychiatry 77(12):1383–1384, PubMed Pubmed Central PMCID: PMC2077408. Epub 2006/11/18. eng
35. Perrin F, Maquet P, Peigneux P, Ruby P, Degueldre C, Balteau E et al (2005) Neural mechanisms involved in the detection of our first name: a combined ERPs and PET study. Neuropsychologia 43(1):12–19, Epub 2004/10/19. eng
36. Thibaut A, Bruno MA, Chatelle C, Gosseries O, Vanhaudenhuyse A, Demertzi A et al (2012) Metabolic activity in external and internal awareness networks in severely brain-damaged patients. J Rehabil Med 44(6):487–494, PubMed Epub 2012/03/01. eng
37. Phillips CL, Bruno MA, Maquet P, Boly M, Noirhomme Q, Schnakers C et al (2011) "Relevance vector machine" consciousness classifier applied to cerebral metabolism of vegetative and locked-in patients. Neuroimage 56(2):797–808, PubMed Epub 2010/06/24. eng
38. Stender J, Gosseries O, Bruno MA, Charland-Verville V, Vanhaudenhuyse A, Demertzi A et al (2014) Diagnostic precision of PET imaging and functional MRI in disorders of consciousness: a clinical validation study. Lancet 384(9942):514–522, PubMed Epub 2014/04/22. Eng
39. Bardin JC, Fins JJ, Katz DI, Hersh J, Heier LA, Tabelow K et al (2011) Dissociations between behavioural and functional magnetic resonance imaging-based evaluations of cognitive function after brain injury. Brain 134(Pt 3):769–782, PubMed Pubmed Central PMCID: PMC3044833. Epub 2011/03/01. eng
40. Biswal BB, Van Kylen J, Hyde JS (1997) Simultaneous assessment of flow and BOLD signals in resting-state functional connectivity maps. NMR Biomed 10(4-5):165–170, PubMed Epub 1997/06/01. eng

41. Beckmann CF, DeLuca M, Devlin JT, Smith SM (2005) Investigations into resting-state connectivity using independent component analysis. Philos Trans R Soc Lond B Biol Sci 360(1457):1001–1013, PubMed Pubmed Central PMCID: PMC1854918. Epub 2005/08/10. eng
42. Raichle ME, MacLeod AM, Snyder AZ, Powers WJ, Gusnard DA, Shulman GL (2001) A default mode of brain function. Proc Natl Acad Sci U S A 98(2):676–682, PubMed Pubmed Central PMCID: PMC14647. Epub 2001/02/24. eng
43. Schilbach L, Eickhoff SB, Rotarska-Jagiela A, Fink GR, Vogeley K (2008) Minds at rest? Social cognition as the default mode of cognizing and its putative relationship to the "default system" of the brain. Conscious Cogn 17(2):457–467, PubMed Epub 2008/04/25. eng
44. Vanhaudenhuyse A, Noirhomme Q, Tshibanda LJ, Bruno MA, Boveroux P, Schnakers C et al (2010) Default network connectivity reflects the level of consciousness in non-communicative brain-damaged patients. Brain 133(Pt 1):161–171, PubMed Pubmed Central PMCID: PMC2801329. Epub 2009/12/26. eng
45. Boveroux P, Vanhaudenhuyse A, Bruno MA, Noirhomme Q, Lauwick S, Luxen A et al (2010) Breakdown of within- and between-network resting state functional magnetic resonance imaging connectivity during propofol-induced loss of consciousness. Anesthesiology 113(5):1038–1053, PubMed Epub 2010/10/05. eng
46. Greicius MD, Kiviniemi V, Tervonen O, Vainionpaa V, Alahuhta S, Reiss AL et al (2008) Persistent default-mode network connectivity during light sedation. Hum Brain Mapp 29(7):839–847, Pubmed Central PMCID: PMC2580760. Epub 2008/01/26. eng
47. Bruno MA, Laureys S, Demertzi A (2013) Coma and disorders of consciousness. Handb Clin Neurol 118:205–213, PubMed Epub 2013/11/05. eng
48. Di Perri C, Bastianello S, Bartsch AJ, Pistarini C, Maggioni G, Magrassi L et al (2013) Limbic hyperconnectivity in the vegetative state. Neurology 81(16):1417–1424, PubMed Epub 2013/09/21. eng
49. Lu H, Zou Q, Gu H, Raichle ME, Stein EA, Yang Y (2012) Rat brains also have a default mode network. Proc Natl Acad Sci U S A 109(10):3979–3984, PubMed Pubmed Central PMCID: PMC3309754. Epub 2012/02/23. eng
50. Boly M, Tshibanda L, Vanhaudenhuyse A, Noirhomme Q, Schnakers C, Ledoux D et al (2009) Functional connectivity in the default network during resting state is preserved in a vegetative but not in a brain dead patient. Hum Brain Mapp 30(8):2393–2400, PubMed Epub 2009/04/08. eng
51. Vincent JL, Patel GH, Fox MD, Snyder AZ, Baker JT, Van Essen DC et al (2007) Intrinsic functional architecture in the anaesthetized monkey brain. Nature 447(7140):83–86, PubMed Epub 2007/05/04. eng
52. Barttfeld P, Uhrig L, Sitt JD, Sigman M, Jarraya B, Dehaene S (2015) Signature of consciousness in the dynamics of resting-state brain activity. Proc Natl Acad Sci U S A 112(3):887–892, PubMed Epub 2015/01/07. eng
53. Heine L, Soddu A, Gomez F, Vanhaudenhuyse A, Tshibanda L, Thonnard M et al (2012) Resting state networks and consciousness: alterations of multiple resting state network connectivity in physiological, pharmacological, and pathological consciousness States. Front Psychol 3:295, PubMed Pubmed Central PMCID: PMC3427917. Epub 2012/09/13. eng
54. Anderson JS, Druzgal TJ, Lopez-Larson M, Jeong EK, Desai K, Yurgelun-Todd D (2011) Network anticorrelations, global regression, and phase-shifted soft tissue correction. Hum Brain Mapp 32(6):919–934, PubMed Pubmed Central PMCID: PMC3220164. Epub 2010/06/10. eng
55. Chai XJ, Castanon AN, Ongur D, Whitfield-Gabrieli S (2012) Anticorrelations in resting state networks without global signal regression. Neuroimage 59(2):1420–1428, PubMed Pubmed Central PMCID: PMC3230748. Epub 2011/09/06. eng
56. Samann PG, Wehrle R, Hoehn D, Spoormaker VI, Peters H, Tully C et al (2011) Development of the brain's default mode network from wakefulness to slow wave sleep. Cereb Cortex 21(9):2082–2093, PubMed Epub 2011/02/19. eng
57. Hoelper BM, Soldner F, Chone L, Wallenfang T (2000) Effect of intracerebral lesions detected in early MRI on outcome after acute brain injury. Acta Neurochir Suppl 76:265–267, PubMed Epub 2001/07/14. eng
58. Kampfl A, Schmutzhard E, Franz G, Pfausler B, Haring HP, Ulmer H et al (1998) Prediction of recovery from post-traumatic vegetative state with cerebral magnetic-resonance imaging. Lancet 351(9118):1763–1767, PubMed Epub 1998/06/23. eng

59. Huisman TA, Loenneker T, Barta G, Bellemann ME, Hennig J, Fischer JE et al (2006) Quantitative diffusion tensor MR imaging of the brain: field strength related variance of apparent diffusion coefficient (ADC) and fractional anisotropy (FA) scalars. Eur Radiol 16(8):1651–1658, PubMed Epub 2006/03/15. eng
60. Jones DK, Dardis R, Ervine M, Horsfield MA, Jeffree M, Simmons A et al (2000) Cluster analysis of diffusion tensor magnetic resonance images in human head injury. Neurosurgery 47(2):306–313; discussion 13-4. PubMed Epub 2000/08/15. eng
61. Huisman TA, Sorensen AG, Hergan K, Gonzalez RG, Schaefer PW (2003) Diffusion-weighted imaging for the evaluation of diffuse axonal injury in closed head injury. J Comput Assist Tomogr 27(1):5–11, PubMed Epub 2003/01/25. eng
62. Arfanakis K, Hermann BP, Rogers BP, Carew JD, Seidenberg M, Meyerand ME (2002) Diffusion tensor MRI in temporal lobe epilepsy. Magn Reson Imaging 20(7):511–519, PubMed Epub 2002/11/05. eng
63. Rutgers DR, Fillard P, Paradot G, Tadie M, Lasjaunias P, Ducreux D (2008) Diffusion tensor imaging characteristics of the corpus callosum in mild, moderate, and severe traumatic brain injury. AJNR Am J Neuroradiol 29(9):1730–1735, PubMed Epub 2008/07/12. eng
64. Nakayama N, Okumura A, Shinoda J, Yasokawa YT, Miwa K, Yoshimura SI et al (2006) Evidence for white matter disruption in traumatic brain injury without macroscopic lesions. J Neurol Neurosurg Psychiatry 77(7):850–855, PubMed Pubmed Central PMCID: PMC2117497. Epub 2006/04/01. eng
65. Kraus MF, Susmaras T, Caughlin BP, Walker CJ, Sweeney JA, Little DM (2007) White matter integrity and cognition in chronic traumatic brain injury: a diffusion tensor imaging study. Brain 130(Pt 10):2508–2519, PubMed Epub 2007/09/18. eng
66. Newcombe V, Chatfield D, Outtrim J, Vowler S, Manktelow A, Cross J et al (2011) Mapping traumatic axonal injury using diffusion tensor imaging: correlations with functional outcome. PLoS One 6(5):e19214, PubMed Pubmed Central PMCID: PMC3087728. Epub 2011/05/17. eng
67. Galanaud D, Perlbarg V, Gupta R, Stevens RD, Sanchez P, Tollard E et al (2012) Assessment of white matter injury and outcome in severe brain trauma: a prospective multicenter cohort. Anesthesiology 117(6):1300–1310, PubMed Epub 2012/11/09. eng
68. Luyt CE, Galanaud D, Perlbarg V, Vanhaudenhuyse A, Stevens RD, Gupta R et al (2012) Diffusion tensor imaging to predict long-term outcome after cardiac arrest: a bicentric pilot study. Anesthesiology 117(6):1311–1321, PubMed Epub 2012/11/09. eng
69. Tollard E, Galanaud D, Perlbarg V, Sanchez-Pena P, Le Fur Y, Abdennour L et al (2009) Experience of diffusion tensor imaging and 1H spectroscopy for outcome prediction in severe traumatic brain injury: preliminary results. Crit Care Med 37(4):1448–1455, PubMed Epub 2009/02/27. eng
70. Gerdes JS, Walther EU, Jaganjac S, Makrigeorgi-Butera M, Meuth SG, Deppe M (2014) Early detection of widespread progressive brain injury after cardiac arrest: a single case DTI and post-mortem histology study. PLoS One 9(3), PubMed Pubmed Central PMCID: PMC3954875. Epub 2014/03/19. eng
71. van der Eerden AW, Khalilzadeh O, Perlbarg V, Dinkel J, Sanchez P, Vos PE et al (2014) White matter changes in comatose survivors of anoxic ischemic encephalopathy and traumatic brain injury: comparative diffusion-tensor imaging study. Radiology 270(2):506–516, PubMed Epub 2014/01/30. eng
72. Wu F, Tang Y, Xu K, Kong L, Sun W, Wang F et al (2011) Whiter matter abnormalities in medication-naive subjects with a single short-duration episode of major depressive disorder. Psychiatry Res 191(1):80–83, PubMed Pubmed Central PMCID: PMC3058813. Epub 2010/12/15. eng
73. Fernandez-Espejo D, Bekinschtein T, Monti MM, Pickard JD, Junque C, Coleman MR et al (2011) Diffusion weighted imaging distinguishes the vegetative state from the minimally conscious state. Neuroimage 54(1):103–112, PubMed Epub 2010/08/24. eng
74. Newcombe VF, Williams GB, Scoffings D, Cross J, Carpenter TA, Pickard JD et al (2010) Aetiological differences in neuroanatomy of the vegetative state: insights from diffusion tensor imaging and functional implications. J Neurol Neurosurg Psychiatry 81(5):552–561, PubMed Epub 2010/05/13. eng

Chapter 6
Decoding Thoughts in Disorders of Consciousness

Adrian M. Owen and Lorina Naci

Abstract Recent developments in functional neuroimaging have provided a number of new tools for assessing patients who clinically appear to be in a vegetative or minimally conscious state. These techniques have been able to reveal awareness and even allow rudimentary communication, in some patients who remain entirely behaviorally nonresponsive. In some centers, these methods are now being employed routinely in the assessment of severely brain-injured patients, mapping patterns of residual function and dysfunction and helping to reduce diagnostic errors in these conditions. The implications of these results extend well beyond the immediate clinical and scientific findings to suggest an urgent need for a reevaluation of the existing diagnostic guidelines for behaviorally nonresponsive patients to include information derived from functional neuroimaging.

Keywords Disorders of consciousness • fMRI • Mental states • Decoding

6.1 Introduction

The definition and assessment of consciousness is still one of the most challenging areas of contemporary neuroscience. Empirically, consciousness is often separated into two components: wakefulness (also known as "arousal," referring to the *level* of consciousness) and awareness (the *contents* of consciousness) [1]. Wakefulness is a state in which the eyes are open and where, typically, there is a degree of motor arousal, while awareness is accompanied by the ability to have (and the act of having) an experience of some kind [2]. Wakefulness is relatively easily measured by confirming that the eyes are open and/or that the resting state electroencephalography (EEG) exhibits a pattern that is typical for the normal waking state. Measuring awareness is much more difficult; indeed, objective functional biomarkers of awareness are lacking (see Supplementary Online Information), and, clinically, the

A.M. Owen (✉) • L. Naci
Department of Psychology, The Brain and Mind Institute, The University of Western Ontario, London, ON N6A 5B7, Canada
e-mail: uwocerc@uwo.ca

M.M. Monti, W.G. Sannita (eds.), *Brain Function and Responsiveness in Disorders of Consciousness*, DOI 10.1007/978-3-319-21425-2_6

ability to follow commands – either verbally or behaviorally – has to be used as a proxy measure for awareness. However, consistent command following is not always possible in severely brain-injured patients and residual awareness has to be inferred on the basis of any behavior that is observed. The subjective nature of this inference undoubtedly contributes to the well-documented rate of misdiagnosis in this population [3–5].

Recent advances in neuroimaging technology have suggested a number of possible solutions to this problem. In this chapter, recent studies that have used functional magnetic resonance imaging (fMRI) in this context will be reviewed.

6.2 Decoding Mental States Using fMRI

In recent years, many attempts have been made to "decode" mental decisions or thoughts in healthy participants using neuroimaging [6–10]. The principle often employed capitalizes on the fact that certain types of thought are associated with a unique brain activation pattern that can be used as a signature for that specific thought. If a statistical classifier is trained to recognize these characteristic signatures, a volunteer's thoughts can be decoded (within the constraints of the experimental design) using their brain activity alone.

To achieve acceptable levels of accuracy, these methods often rely on mental imagery as a proxy for the physical response being decoded. For example, in one early study, four non-naive participants learned, with the aid of feedback, to willfully regulate their fMRI signal using self-chosen visual imagery strategies (e.g., pictures of buildings, spatial navigation, clenching, dancing) [10]. In a more sophisticated design, information derived from both the timing (onset and offset) and the source location of the hemodynamic response was used to decode which of four possible answers was being given to questions [11]. To indicate their choice (or "thought"), participants imagined one of two tasks, beginning at one of four times and continuing for different pre-specified durations. An automated decoding procedure deciphered the answer by analyzing the single-trial BOLD responses in real time with a mean accuracy of 94.9 %.

In one large study [6], 34 healthy volunteers were asked to imagine hitting a tennis ball back and forth to an imaginary coach when they heard the word "tennis" (thereby eliciting vigorous imaginary arm movements) and to imagine walking from room to room in their house when they heard the word "house" (thereby eliciting imaginary spatial navigation). Imagining playing tennis was associated with robust activity in the supplementary motor area in each and every one of the participants scanned. In contrast, imagining moving from room to room in a house activated the parahippocampal cortices, the posterior parietal lobe, and the lateral premotor cortices, all regions that have been shown to contribute to imaginary or real spatial navigation [12]. By simply examining the responses elicited during the imagery tasks, Boly and colleagues [6] were able to "decode" which task was being mentally "performed." Moreover, the robustness and reliability of fMRI responses

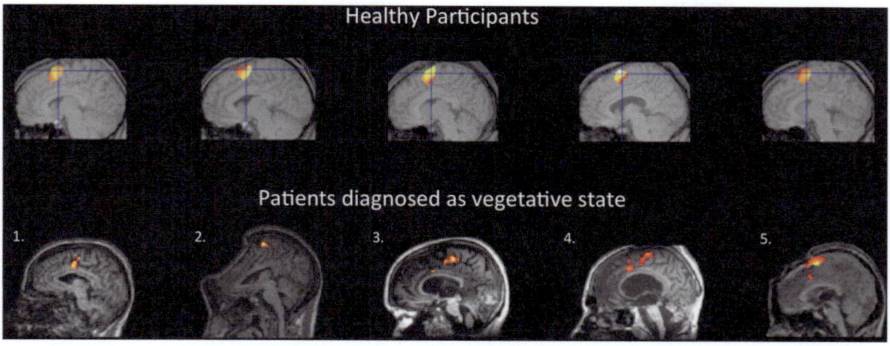

Fig. 6.1 *Top row*: Five healthy participants asked to imagine playing tennis in the fMRI scanner (Adapted from [6]). In all five cases, significant activity was observed in the premotor cortex, indicating that they had understood the instruction and were responding by carrying out the appropriate type of mental imagery, that is, *following a command. Bottom row*: Formally identical responses in five patients who behaviorally meet the clinical criteria for a diagnosis of vegetative state (Adapted from Owen et al. [14] (patient 5) and [15] (patients 1–4)) confirming that, in spite of an inability to respond physically, these patients can still demonstrate *command following* by modulating their cortical fMRI activity. Such responses are observed in approximately 17 % of vegetative patients

across individuals meant that activity in these regions could be used to confirm that the participants retained the ability to understand instructions to carry out different mental tasks in response to those instructions and, therefore, were able to exhibit willed, voluntary brain behavior in the absence of any overt action. On this basis, Boly and colleagues [6] argued that, like any other form of action that requires a choice between one of several possible responses, these brain responses require *awareness*, that is to say, awareness of the various contingencies that govern the relationship between a given stimulus (in this case, the cue word for one of two possible imagery tasks) and a response (in this case, imagining a type of action). Put simply, fMRI responses of this sort can be used to measure awareness because awareness is necessary for them to occur.

Owen et al. [13, 14] used this same logic to demonstrate that a young woman who fulfilled all internationally agreed criteria for the vegetative state was, in fact, consciously aware and able to make responses of this sort using her brain activity. The patient, who was involved in a complex road traffic accident and had sustained very severe traumatic brain injuries, had remained entirely unresponsive for a period of 6 months prior to the fMRI scan. During two different scanning sessions, the patient was instructed to perform the two mental imagery tasks described above. In each case she was asked to imagine playing tennis/moving around the rooms of her home (for 30 s) when she heard the word *tennis/house* and to relax (for 30 s) when she heard the word *relax*. When she was asked to imagine playing tennis (Fig. 6.1, patient 5), significant activity was observed repeatedly in the supplementary motor area [14] that was indistinguishable from that observed in the healthy volunteers scanned by Boly et al. [6]. Moreover, when she was asked to imagine walking through her home, significant activity was observed in the parahippocampal gyrus, the posterior parietal cortex, and the lateral premotor cortex which was again indistinguishable from those observed in healthy volunteers [13, 14]. The patient's brain activity was statistically robust, reproducible, task appropriate (enhanced following the "tennis"/"house" cue and returning to baseline following the "relax" cue), sustained over long time intervals (30 s), and repeated over each 5-min session. On this basis, it was concluded that, despite fulfilling all of the clinical criteria for a diagnosis of vegetative state, this patient retained the ability to understand spoken commands and to respond to them through her brain activity, rather than through speech or movement, confirming that she was consciously aware of herself and her surroundings. In a follow-up study of 23 patients who were behaviorally diagnosed as vegetative, Monti et al. [15] showed that 4 (17 %) were able to generate reliable responses of this sort in the fMRI scanner (Fig. 6.1, patients 1–4).

After a severe brain injury, when the request to move a hand or a finger is followed by an appropriate motor response, the diagnosis can change from vegetative state (no evidence of awareness) to minimally conscious state (some evidence of awareness). By analogy then, if the request to activate, say, the supplementary motor area of the brain by imagining moving the hand is followed by an appropriate brain response, we should give that response the very same weight [16–18]. Skeptics may argue that brain responses are somehow less physical, reliable, or immediate than motor responses but, as is the case with motor responses, all of these arguments can be dispelled with careful measurement, replication, and objective verification [6, 14, 15, 16, 19, 20, 21]. For example, if a patient who was assumed to be unaware raised his or her hand to command on just one occasion, there would remain some doubt about the presence of awareness given the possibility that this movement was a chance occurrence, coincident with the instruction. However, if that same patient were able to repeat this response to command on ten occasions, there would remain little doubt that the patient was aware. By the same token, if that patient was able to activate his or her supplementary motor area in response to command (e.g., by being told to imagine hand movements) and was able to do this on every one of ten trials, we would have to accept that this patient was consciously aware. Like most neuroimaging investigations, replication of this sort was inherent in both of the studies described above [14, 15], because the statistically significant results depended on multiple, similar responses being exhibited across repeated trials.

Monti et al. [15] employed the same general principle – that is, using the neural signature of active mental rehearsal to signify awareness – to show that communication of "yes" and "no" responses was possible with fMRI. Thus, three "yes" and "no" responses were decoded from each of 16 healthy participants with 100 % accuracy using only their real-time changes in the supplementary motor area (during tennis imagery) and the parahippocampal place area (during spatial navigation). Moreover, in one traumatic brain injury patient, who had been repeatedly diagnosed as vegetative over a 5-year period, similar questions were posed and successfully decoded using the same approach [15]. Indeed, this patient was able to convey biographical information that was not known to the experimenters at the time (but was verified as factually correct) such as his father's name and the last place that he had visited on vacation before his accident 5 years earlier. In contrast, and despite a reclassification to minimally conscious state following the fMRI scan, it remained impossible to establish any form of communication with this patient at the bedside.

An obvious application for approaches of this sort is to begin to involve some of these patients in the decision-making processes involved in their own therapeutic care and management. To date, this has only been achieved successfully in one patient, who had been repeatedly diagnosed as vegetative for 12 years following a traumatic brain injury [16]. The patient was a male who, in December 1999 and at the age of 26, had suffered a severe closed head injury in a motor-vehicle accident. On admission to hospital he had a Glasgow Coma Scale [22] score of 4, meaning that he was unable to open his eyes or produce any sound and his only response was extension to painful stimulation. Over the next 12 years, the patient was assessed regularly by experienced neurologists and multidisciplinary teams and throughout this period his behavior remained consistent with the internationally accepted criteria for the vegetative state. Indeed, over one 14-month period in 2011–2013, a total of 20 standardized behavioral assessments were performed by a multidisciplinary team, at different times of the day and in different postural positions, using the Coma Recovery Scale-Revised [23], and his diagnosis was unchanged throughout. In February 2012, 12 years and 2 months after his accident, the patient was first scanned using the fMRI mental imagery approach described above [14, 15]. The patient was able to provide correct answers to multiple externally verifiable questions, including his own name, his whereabouts, the name of his personal support worker (who he had only encountered in the years following his accident), the current date, and other basic factual information (e.g., whether a banana is yellow). Two non-verifiable questions were then posed, including one pertaining to his care preferences (e.g., whether he liked watching (ice) hockey games on TV) and another to details about his current clinical condition (e.g., whether he was in any physical pain). Within the time constraints of the scanning visits, the majority of responses to these questions were verified in independent sessions that posed the reverse questions (e.g., "Is your name Mike?" vs. "Is your name Scott?"). In total, answers to 12 different questions were obtained across several sessions, despite the fact that the patient remained entirely physically non-responsive at the bedside [16].

Although techniques like the ones described above require that the patient engages in rather specific types of mental imagery (playing tennis or moving from room to room through a house), that is not really the main point that allows consciousness to be detected and communication to occur. All that is required to detect consciousness is a reliable indicator that a patient can turn his or her attention to a specific scenario, because this then serves as a "neural proxy" for a physical "response to command." By extension, if it can be shown that the patient can turn his or her attention to two separate scenarios, then communication is possible because those two separate scenarios can be linked to "yes" responses and "no" responses, respectively. Thus, mental imagery is not necessary at all, but serves as a simple vehicle for guiding a patient's attention one way or another.

A related and possibly simpler approach to detecting covert awareness after brain injury, therefore, is to target processes that require the willful adoption of "mind-sets" in carefully matched (perceptually identical) experimental and control conditions. For example, Monti et al. [24] presented healthy volunteers with a series of neutral words and alternatively instructed them to just listen, or to count, the number of times a given word was repeated. As predicted, the counting task revealed the frontoparietal network that has been previously associated with target detection and working memory. When tested on this same procedure, a severely brain-injured patient produced a very similar pattern of activity, confirming that he could willfully adopt differential mind-sets as a function of the task conditions and could actively maintain these mind-sets across time. These covert abilities were entirely absent from his documented behavioral repertoire. As in the tennis/spatial navigation examples described above, because the external stimuli (a series of words) were identical in the two conditions, any difference in brain activity observed cannot reflect an "automatic" brain response (i.e., one that can occur in the absence of consciousness). Rather, the activity must reflect the fact that the patient has performed a particular action (albeit a "brain action") in response to the stimuli on one (but not the other) presentation; in this sense, the brain response is entirely analogous to a (motor) response to command and should carry the same weight as evidence of awareness.

Naci and colleagues [20, 21] took this general principle even further and developed a novel tool for communicating with nonresponsive patients based on how they selectively directed their attention to sounds while in the fMRI scanner. It is well established that selective attention can significantly enhance the neural representation of attended sounds [25], although most previous studies have focused on group-level changes rather than individual responses that are crucial for work with (individual) brain-injured patients. In their first study [20], 15 healthy volunteers answered questions (e.g., "Do you have brothers or sisters?") in the fMRI scanner, by selectively attending to the appropriate word ("yes" or "no"), which was played to them auditorily, interspersed with "distractor" stimuli (digits 1–9). Ninety percent of the answers were decoded correctly based on activity changes within the attention network of the brain. Moreover, the majority of volunteers conveyed their answers with less than 3 min of scanning, which represents a significant time saving over the mental imagery methods described above [6, 14]. Indeed, a formal comparison between the two approaches revealed improved individual success rates and an overall reduction in the scanning times required to correctly detect responses; 100 % of volunteers showed significant task-appropriate activity to the selective attention task, compared to 87 % to the motor imagery. This result is consistent with previous studies showing that a proportion of healthy volunteers do not produce reliable brain activation during mental imagery tasks [6].

In a follow-up study, Naci and Owen [21] used the same approach to test for residual conscious awareness and communication abilities in three behaviorally nonresponsive, brain-injured patients. As in the previous study of healthy participants, the patients had to either "count" or "relax" as they heard a sequence of sounds. The word *count* at the beginning of the sequence instructed the patient to count the occurrences of a target word (*yes* or *no*), while the word *relax* instructed them to relax and ignore the sequence of words. Reliable activity increases in the attention network of the brain after the word *count* relative to the word *relax* was taken as evidence of command following. All three patients (two of whom were diagnosed as being in a minimally conscious state and one as being in a vegetative state) were able to convey their ability to follow commands inside the fMRI scanner by following the instructions in this way. In stark contrast, extremely limited or a complete lack of behavioral responsivity was observed in repeated bedside assessments of all three patients. These results confirm that selective attention is an appropriate vehicle for detecting covert awareness in some behaviorally nonresponsive patients who are presumed to mostly or entirely lack any cognitive abilities whatsoever.

In a following series of scans, communication was attempted in two of the patients. The communication scans were similar to those in the command-following scan, with one exception. Instead of an instruction (count or relax), a binary question (e.g., "Is your name Steven?") preceded each sound sequence. Thus, each patient then had to willfully choose which word to attend to (count) and which to ignore, depending on which answer he wished to convey to the specific question that had been asked. Using this method, the two patients (one diagnosed as minimally conscious state and one diagnosed as vegetative state) were able to use selective attention to repeatedly communicate correct answers to questions that were posed to them by the experimenters [21]. In the absence of external cues as to which word the patient was attending to, the functional brain activation served as the only indicator of the patient's intentions and, in both cases, led to the correct answers being decoded. For example, when asked, "Are you in a supermarket?" one patient showed significantly more activation for "no" than "yes" sequences in a network of brain areas that had been previously activated when that patient was focusing attention on external cues (Fig. 6.2). Conversely, when asked, "Are you in a hospital?," the patient showed significantly more activation for "yes" than "no" sequences in the same regions. Despite his diagnosis (vegetative state for 12 years), the fMRI approach allowed this patient to establish interactive communication with the research team in 4 different fMRI sessions. The patient's brain responses within specific regions were remarkably consistent and reliable across two different scanning visits, 5 months apart, during which the patient maintained the long-standing vegetative state diagnosis. For all four questions, the patient produced a robust neural response and was able to provide the correct answer with 100 % accuracy. The patient's brain activity in the communication scans not only further corroborated that he was, indeed, consciously aware but also revealed that he had far richer cognitive reserves than could be assumed based on his clinical diagnosis. In particular, beyond the ability to pay attention, these included autobiographical knowledge and awareness of his location in time and space.

These types of approaches all illustrate a paradigmatic shift toward the use of active (e.g., willful) tasks in the assessment of covert awareness after serious brain injury. What sets such tasks apart is that the neural responses required are not produced *automatically* by the eliciting stimulus, but rather, depend on time-dependent

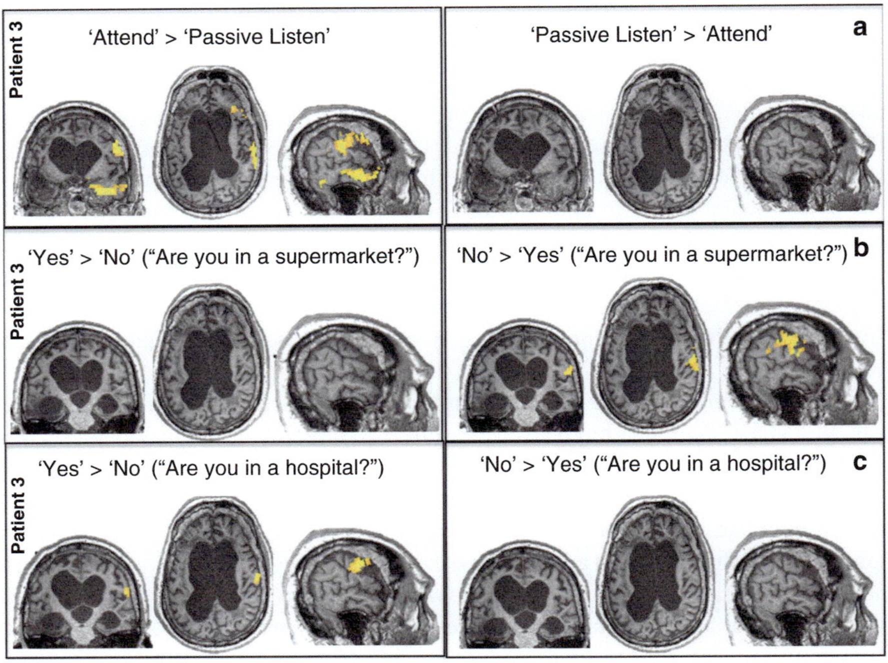

Fig. 6.2 Command-following (**a**) and communication (**b**, **c**) scans in patient 3, clinically diagnosed as being in a vegetative state. Brain activity is overlaid on the patient's native anatomic volume. The opposite directions of each contrast (i.e., $a>b$ or $b>a$) are shown on the left and right sides of each panel. (**a**) The command-following scan also served to localize the brain foci of attention unique to the patient. (**b**, **c**) Selective attention to the answer word (either yes or no) during each communication scan was investigated within these regions. Attention to the answer in each question (**b**, no; **c**, yes) significantly activated the precentral or motor region

and sustained responses generated by the participants themselves. Such behavior (albeit neural "behavior") provides a proxy for a motor action and is, therefore, an appropriate vehicle for reportable awareness [26].

Nevertheless, while "active" paradigms have proven themselves to be an effective means for assessing residual awareness in some nonresponsive patients, it remains likely that many patients will lack the necessary cognitive resources for carrying out these tasks in the scanner and will therefore fail to exhibit signs of awareness even when it may exist. To address this issue, recent efforts have focused on developing new methods for detecting awareness in the absence of an explicitly willed task. Naci and colleagues [27] used a richly evocative stimulus – a highly suspenseful movie – to capture attention naturally in the absence of structured instruction. They asked whether a common neural basis can account for how different individuals form similar conscious experiences and, if so, whether it could be used to interpret those experiences without recourse to self-report in behaviorally nonresponsive patients. They reasoned that executive function, in particular, might provide an empirical window by which the cognitive aspect of human conscious experience can be quantified. By their very nature, engaging movies are designed to give viewers a shared conscious experience driven, in part, by the recruitment of similar executive processes, as each viewer continuously integrates their observations, analyses, and predictions while filtering out any distractions, leading to an ongoing involvement in the movie's plot.

When healthy participants viewed a highly engaging short movie by Alfred Hitchcock – the so-called Master of Suspense – in the fMRI scanner, they displayed highly synchronized brain activity in supramodal frontal and parietal regions, which support executive function [28, 29]. The movie's executive demands, assessed quantitatively with a dual-task procedure [30] by an independent group, predicted activity in frontal and parietal regions of the healthy participants, who had watched the movie without a secondary task in the scanner. Importantly, the movie's suspense ratings, provided by a third independent healthy group, demonstrated that individual participants had a similar qualitative experience of the movie, which also predicted activity in the frontal and parietal regions. Together, these results suggested that the movie's executive demands drove brain activity in frontal and parietal regions and, further, that the synchronization of this activity across individuals underpinned their similar experience. By extension, the degree to which each individual's frontoparietal brain activity could be predicted from the rest of the group's represented a reliable neural index of how similar his or her cognitive experience was to the others'.

Naci et al. [27] then used this approach in two entirely behaviorally nonresponsive patients with unknown levels of consciousness, in order to examine and quantify their experience of the world. fMRI data was acquired from the two patients, as they freely viewed the same Hitchcock movie. One patient, who had remained behaviorally nonresponsive for a 16-year period prior to scanning, demonstrated a highly similar brain response to that of the three independent control groups. The patient's brain activity in frontal and parietal regions was tightly synchronized with the healthy participants' over time, and crucially, it reflected the executive demands of specific events in the movie, as measured both quantitatively and qualitatively in healthy individuals (Fig. 6.3). This suggested that the patient could continuously engage in complex thoughts about real-world events unfolding over time and, thus, that he was consciously aware. Further, the patient's brain response suggested that his conscious experience was highly similar to that of each and every healthy participant, including his moment-to-moment perception of the movie content, as well as his executive engagement with its plot. These processes are likely to include updating the contents of working memory (e.g., to follow the plot), relating events in the movie to past experiences (e.g., to appreciate that a gun is a dangerous weapon), and coding the foreshadowing cues (i.e., events that might have future relevance to the plot) characteristic of movies of this type.

In summary, this approach can determine not only whether any given patient is conscious but also infer what the contents of that conscious experience might actually be, thus, revealing important practical and ethical implications for the patient's standard of care and quality of life [31].

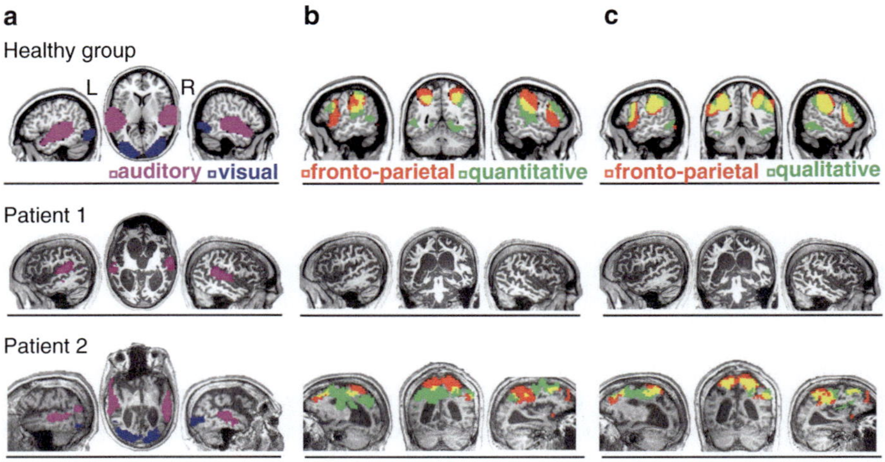

Fig. 6.3 Decoding executive function in one behaviorally nonresponsive patient. *Healthy group*: (**a**) group-level auditory (*purple*) and visual (*blue*) independent components. (**b**, **c**) The healthy group's activity predicted by the quantitative (**b**)/qualitative (**c**) executive measure (*green*) is overlaid on the group frontoparietal independent component (*red*); overlap areas are displayed in yellow. *Patient 1*: (**a**) The healthy group's auditory independent component predicted significant activity in patient 1's auditory cortex (*purple*). (**b**, **c**) No evidence of visual responses or executive processing similar to the healthy participants' was observed. *Patient 2*: (**a**) The healthy group's auditory and visual independent components predicted significant activity in patient 2's auditory (*purple*) and visual (*blue*) cortex, respectively. (**b**, **c**) The quantitative (**b**) and qualitative (**c**) executive measures predicted activity (*green*) in the patient's frontal and parietal regions. Overlap with activity predicted by the healthy group's frontoparietal independent component (*red*) is displayed in *yellow* (Adapted from [27])

6.3 Diagnostic Implications

An obvious clinical consequence of the emergence of novel neuroimaging techniques that permit the identification of covert awareness and communication in the absence of any behavioral response is the possibility of improved diagnosis after severe brain injury. It is notable that in one of the cases described above [16, 21, 32], the patient was repeatedly and rigorously assessed by experienced teams and showed no behavioral sign of awareness on any of these occasions – indeed, this continued to be the case even after awareness had been established unequivocally with both fMRI and EEG. Technically however, he was not *misdiagnosed* (as vegetative), in the sense that any error of judgment was made, because the accepted diagnostic criteria are based on behavior, and no behavioral marker of awareness was missed. Nevertheless, the existing criteria did not accurately capture his actual state of awareness and, in this sense, his vegetative state diagnosis was clearly incorrect. What then is the appropriate diagnostic label for such patients, and who can follow commands with a measurable brain response, but physically remain entirely nonresponsive? The term "non-behavioral minimally conscious state" has been suggested [33], although because attention, language comprehension, and working memory are demonstrably preserved in these patients, we have argued that "minimally conscious" does not adequately describe their residual cognitive abilities [16, 17, 21]. Indeed, the patient described above was consistently and reliably able to communicate (using fMRI), which places him well beyond the diagnostic criteria describing the minimally conscious state. The term "functional locked-in syndrome" has also been proposed for patients who demonstrate consistent and reliable communication using solely adjunctive technologies [34, 35]. In its classical clinical presentation, "locked-in syndrome" refers to patients who are left with only vertical eye movements and/or blinking, which often permits rudimentary communication. Cognitive function, however, is generally fully preserved, at least in those cases where the lesion is limited to the ventral pons [36]. Patients like the one described here are clearly "locked in" in the general sense of the term, but do not have many of the same neuropathological and clinical features of the classic locked-in syndrome. Moreover, at present there is still considerable uncertainty about the full extent of residual cognitive function in such patients, and thus about the suitability of the term "functional locked-in syndrome." That said, this is precisely the sort of question that can be explored with fMRI. Indeed, the patient has already been able to report that he remembers his own name and that he knows the current date and where he is [21], confirming that he is well oriented in time and space. He has also provided information about events that have occurred in the years since his accident, confirming that he is still able to encode new memories. Schnakers et al. [36] have recently developed a standardized neuropsychological assessment for locked-in syndrome that uses simple eye movements as responses (in most cases to provide "yes"/"no" answers to questions). There is no technical or theoretical reason why a similar approach could not be used with fMRI data in entirely nonresponsive patients, although the data would take considerably longer to acquire. To this end, Hampshire et al. [19] have recently used fMRI to assess complex logical reasoning ability in a patient who was assumed to be in a vegetative state. Adapting a verbal reasoning paradigm from Baddeley [37], Hampshire et al. [19] presented participants with statements describing the ordering of two objects, a face and a house. Participants were instructed to deduce which of the objects was in front and to visualize the object in their mind. For example, if they heard the statement "the face is not followed by a house," the correct answer would be "house." Conversely, if they heard "the face precedes the house," the correct answer would be "face." One patient, who based on the behavioral diagnosis was assumed to be in the vegetative state, engaged the same brain regions as healthy individuals in response to the reasoning task demands. This result was consistent with the patient's positive outcome in the fMRI command following task [6, 14] and suggested that, despite the long-standing clinical diagnosis of vegetative state, the patient was not only consciously aware, but, critically, retained capacity for higher-order cognition, in particular, for solving logically complex verbal problems.

6.4 False-Negative Results

False-negative findings in functional neuroimaging studies are common, even in healthy volunteers, and they present particular difficulties in this patient population. For example, a patient may fall asleep during the scan or may not have properly heard or understood the task instructions, leading to an erroneous negative result. Indeed, in the recent study by Monti et al. [15], no willful fMRI responses were observed in 19 of 23 patients – whether these are *true*-negative findings (i.e., those 19 patients were indeed vegetative) or *false-negative* findings (i.e., some of those patients were conscious, but this was not detected on the day of the scan) cannot be determined. Accordingly, negative fMRI and EEG findings in patients should never be used as evidence for impaired cognitive function or lack of awareness.

6.5 Conclusions

In the last few years, neuroimaging methods have been brought to bear on one of the most complex and challenging questions in clinical medicine, that of detecting residual cognitive function, and even covert awareness, in patients who have sustained severe brain injuries. The results demonstrate that responses need no longer be *physical* responses in the traditional sense (e.g., the blink of an eye or the squeezing of a hand), but can now include responses that occur entirely within the brain itself. The recent use of reproducible and robust task-dependent fMRI responses as a form of "communication" in patients who are assumed to be vegetative [15, 16, 21] represents an important milestone in this process. In some cases, these patients have been able to communicate information that was not known by the experimenters at the time, yet could be independently verified later (using more traditional methods of communication with the family), as being factually correct and true [15, 21]. More importantly perhaps, in one case, a patient has used these methods to answer clinically and therapeutically relevant questions (including "Are you in any pain?") that could not be answered in any other way, including via third party. Findings such as these have profound implications for clinical care, diagnosis, prognosis, and end-of-life decision-making but also shed light on more basic scientific questions about the nature of conscious behavior and the neural representation of our own thoughts and intentions.

Acknowledgments This work was funded by generous awards to Adrian M. Owen from the Canada Excellence Research Chair (CERC) Program, the Canadian Institutes of Health Research (CIHR), and the Natural Sciences and Engineering Research Council of Canada (NSERC).

References

1. Plum F, Posner JB (1982) The diagnosis of stupor and coma. Oxford University Press, Oxford, Uk
2. Royal College of Physicians (2003) The vegetative state: guidance on diagnosis and management. Clin Med 3(3):249–254
3. Andrews K, Murphy L, Munday R, Littlewood C (1996) Misdiagnosis of the vegetative state: retrospective study in a rehabilitation unit. BMJ 313(7048):13–16
4. Childs NL, Mercer WN, Childs HW (1993) Accuracy of diagnosis of persistent vegetative state. Neurology 43(8):1465–1467

5. Schnakers C, Vanhaudenhuyse A, Giacino J, Ventura M, Boly M, Majerus S et al (2009) Diagnostic accuracy of the vegetative and minimally conscious state: clinical consensus versus standardized neurobehavioral assessment. BMC Neurol 9:35. doi:10.1186/1471-2377-9-35
6. Boly M, Coleman MR, Davis MH, Hampshire A, Bor D, Moonen G et al (2007) When thoughts become action: an fMRI paradigm to study volitional brain activity in non-communicative brain injured patients. Neuroimage 36(3):979–992. doi:10.1016/j.neuroimage.2007.02.047
7. Cerf M, Thiruvengadam N, Mormann F, Kraskov A, Quiroga RQ, Koch C et al (2010) On-line, voluntary control of human temporal lobe neurons. Nature 467(7319):1104–1108. doi:10.1038/nature09510
8. Gallivan JP, McLean DA, Valyear KF, Pettypiece CE, Culham JC (2011) Decoding action intentions from preparatory brain activity in human parieto-frontal networks. J Neurosci 31(26):9599–9610. doi:10.1523/JNEUROSCI.0080-11.2011
9. Haynes JD, Sakai K, Rees G, Gilbert S, Frith C, Passingham RE (2007) Reading hidden intentions in the human brain. Curr Biol 17(4):323–328. doi:10.1016/j.cub.2006.11.072
10. Weiskopf N, Mathiak K, Bock SW, Scharnowski F, Veit R, Grodd W et al (2004) Principles of a brain-computer interface (BCI) based on real-time functional magnetic resonance imaging (fMRI). IEEE Trans Biomed Eng 51(6):966–970. doi:10.1109/TBME.2004.827063
11. Sorger B, Dahmen B, Reithler J, Gosseries O, Maudoux A, Laureys S et al (2009) Another kind of 'BOLD Response': answering multiple-choice questions via online decoded single-trial brain signals. Prog Brain Res 177(275–92):275–292. doi:10.1016/S0079-6123(09)17719-1
12. Aguirre GK, Detre JA, Alsop DC, D'Esposito M (1996) The parahippocampus subserves topographical learning in man. Cereb Cortex 6(6):823–829
13. Owen AM, Coleman MR, Boly M, Davis MH, Laureys S, Jolles D et al (2007) Response to comments on "Detecting awareness in the vegetative state". Science 315(5816). doi:10.1126/science.1135583
14. Owen AM, Coleman MR, Boly M, Davis MH, Laureys S, Pickard JD (2006) Detecting awareness in the vegetative state. Science 313(5792):1402
15. Monti MM, Vanhaudenhuyse A, Coleman MR, Boly M, Pickard JD, Tshibanda L et al (2010) Willful modulation of brain activity in disorders of consciousness. N Engl J Med 362(7):579–589. doi:10.1056/NEJMoa0905370
16. Fernandez-Espejo D, Owen AM (2013) Detecting awareness after severe brain injury. Nat Rev Neurosci 14(11):801–809. doi:10.1038/nrn3608
17. Owen AM (2013) Detecting consciousness: a unique role for neuroimaging. Annu Rev Psychol 64:109–133. doi:10.1146/annurev-psych-113011-143729
18. Owen AM, Coleman MR (2008) Functional neuroimaging of the vegetative state. Nat Rev Neurosci 9(3):235–243. doi:10.1038/nrn2330
19. Hampshire A, Parkin BL, Cusack R, Espejo DF, Allanson J, Kamau E et al (2012) Assessing residual reasoning ability in overtly non-communicative patients using fMRI. Neuroimage Clin 2:174–183. doi:10.1016/j.nicl.2012.11.008
20. Naci L, Cusack R, Jia VZ, Owen AM (2013) The brain's silent messenger: using selective attention to decode human thought for brain-based communication. J Neurosci 33(22):9385–9393. doi:10.1523/JNEUROSCI.5577-12.2013
21. Naci L, Owen AM (2013) Making every word count for nonresponsive patients. JAMA Neurol 70(10):1235–1241. doi:10.1001/jamaneurol.2013.3686
22. Teasdale G, Jennett B (1974) Assessment of coma and impaired consciousness A practical scale. Lancet 2(7872):81–84
23. Giacino JT, Kalmar K, Whyte J (2004) The JFK coma recovery scale-revised: measurement characteristics and diagnostic utility. Arch Phys Med Rehabil 85(12):2020–2029
24. Monti MM, Coleman MR, Owen AM (2009) Executive functions in the absence of behavior: functional imaging of the minimally conscious state. Prog Brain Res 177:249–260. doi:10.1016/S0079-6123(09)17717-8
25. Bidet-Caulet A, Fischer C, Besle J, Aguera PE, Giard MH, Bertrand O (2007) Effects of selective attention on the electrophysiological representation of concurrent sounds in the human auditory cortex. J Neurosci 27(35):9252–9261. doi:10.1523/JNEUROSCI.1402-07.2007

26. Zeman A (2009) The problem of unreportable awareness. Prog Brain Res 177:1–9. doi:10.1016/S0079-6123(09)17701-4
27. Naci L, Cusack R, Anello M, Owen AM (2014) A common neural code for similar conscious experiences in different individuals. Proc Natl Acad Sci U S A 111(39):14277–14282. doi:10.1073/pnas.1407007111
28. Barbey AK, Colom R, Solomon J, Krueger F, Forbes C, Grafman J (2012) An integrative architecture for general intelligence and executive function revealed by lesion mapping. Brain 135(Pt 4):1154–1164. doi:10.1093/brain/aws021
29. Duncan J (2010) The multiple-demand (MD) system of the primate brain: mental programs for intelligent behaviour. Trends Cogn Sci 14(4):172–179. doi:10.1016/j.tics.2010.01.004
30. Robertson IH, Manly T, Andrade J, Baddeley BT, Yiend J (1997) 'Oops!': performance correlates of everyday attentional failures in traumatic brain injured and normal subjects. Neuropsychologia 35(6):747–758
31. Peterson A, Naci L, Weijer C, Cruse D, Fernández-Espejo D, Graham M et al (2013) Assessing decision-making capacity in the behaviorally nonresponsive patient with residual covert awareness. AJOB Neuroscience 4(4):3–14
32. Cruse D, Chennu S, Fernandez-Espejo D, Payne WL, Young GB, Owen AM (2012) Detecting awareness in the vegetative state: electroencephalographic evidence for attempted movements to command. PLoS One 7(11), e49933. doi:10.1371/journal.pone.0049933
33. Fins JJ, Schiff ND (2006) Shades of gray: new insights into the vegetative state. Hastings Cent Rep 36(6):8
34. Giacino JT, Schnakers C, Rodriguez-Moreno D, Kalmar K, Schiff N, Hirsch J (2009) Behavioral assessment in patients with disorders of consciousness: gold standard or fool's gold? Prog Brain Res 177:33–48. doi:10.1016/S0079-6123(09)17704-X
35. Laureys S, Schiff ND (2012) Coma and consciousness: paradigms (re)framed by neuroimaging. Neuroimage 61(2):478–491. doi:10.1016/j.neuroimage.2011.12.041
36. Schnakers C, Majerus S, Goldman S, Boly M, Van Eeckhout P, Gay S et al (2008) Cognitive function in the locked-in syndrome. J Neurol 255(3):323–330. doi:10.1007/s00415-008-0544-0
37. Baddeley AD (1968) A 3 min reasoning test based on grammatical transformation. Psychon Sci 10(10):341–342

Chapter 7
Brain Responsiveness After Severe Brain Injury: Revolutions and Controversies

Evan S. Lutkenhoff and Martin M. Monti

Abstract In the past 18 years, neuroimaging techniques have become central to studying the healthy and pathological brain. In the context of disorders of consciousness, neuroimaging approaches have radically changed our understanding of how much cortical and cognitive function can be retained after severe brain injury. In this chapter we review some of the main contributions these approaches have given to the field of disorders of consciousness, and we discuss both the power of these techniques and the main complexities tied to employing them to assess residual brain function in otherwise nonresponsive individuals.

Keywords Vegetative state • Minimally conscious state • Functional MRI • Neuroimaging

7.1 Introduction

Disorders of consciousness (DOC) such as coma, the vegetative state (VS), and the minimally conscious state (MCS), are a group of perplexing conditions of the human brain [1]. Acquired after severe traumatic or nontraumatic brain injury (BI), these conditions represent a spectrum along the continuous space described by the two cardinal elements of consciousness: wakefulness (i.e., the *level* of consciousness) and awareness (i.e., the *contents* of consciousness) [2, 3].

E.S. Lutkenhoff
Department of Psychology, University of California Los Angeles,
Los Angeles, CA 90095, USA

M.M. Monti (✉)
Department of Psychology, University of California Los Angeles,
Los Angeles, CA 90095, USA

Department of Neurosurgery, Brain Injury Research Center, Geffen School of Medicine
at University of California, Los Angeles, Los Angeles, CA 90095, USA
e-mail: monti@ucla.edu

M.M. Monti, W.G. Sannita (eds.), *Brain Function and Responsiveness in Disorders of Consciousness*, DOI 10.1007/978-3-319-21425-2_7

A natural framework for studying how consciousness is generated by the interactions of billions of neurons, the past 18 years have seen the flourishing of studies aimed at understanding the changes in neural function and structure that underlie the loss and (sometimes) recovery of consciousness after severe BI (see Monti [4] for a recent comprehensive review).

In what follows, we take stock of almost 20 years of neuroimaging studies in DOC discussing the main contributions that these methods have brought to the field, as well as the problems and complexities tied to the use of these methods as a means of estimating the degree of residual cognitive functioning, and consciousness, available in DOC patients.

7.2 Unconscious but Not Apallic

As originally conceived, the "vegetative state" label was employed to describe patients who, "recover[ed] from sleep-like coma," but in the "absence of function in the cerebral cortex as judged behaviorally" [5]. Borrowing from the Oxford English dictionary, the word "vegetative" was used to describe an organism capable of remaining alive, but incapable of generating *sensation* and *thought*, and thus devoid of intellectual ability or social intercourse. Although recognized as a problematic terminology by Jennett and Plum, the idea of a brain with preserved wakefulness in the absence of any cortical function had already been captured, in the early 1940s, with the label apallic (from the Latin *a-pallium,* "without a cortex"), which referred to a state of "complete loss of higher (telencephalic) function with an isoelectric encephalography (EEG) and much-reduced cerebral blood-flow and metabolism in supratentorial structures" [5].

Adding up the existing neuroimaging evidence to date, the idea of a silent cerebrum devoid of cortical function has been overwhelmingly shown to be incorrect. In a landmark study of residual brain function after severe BI, Menon and colleagues presented the case of a 26-year-old female who lost consciousness after an acute febrile illness of unknown origin, which eventually culminated in loss of consciousness and a (persistent) VS diagnosis [6]. When the patient was placed in a positron emission tomography (PET) machine and presented with pictures of familiar faces, she exhibited a significant metabolic response, within early and higher-level visual cortices, comparable to that observed in healthy volunteers. Since then, a large number of PET, functional magnetic resonance imaging (fMRI), and EEG studies have definitely put to rest the idea that VS patients are apallic or, in any other sense, devoid of cortical function. Neural responses comparable to those seen in healthy volunteers have since been observed in the context of visual [7, 8], auditory [9, 10], linguistic [11, 12], and noxious [13, 14] stimulation, as well as simple forms of learning [15, 16]. So, unconscious, yes, but certainly not *apallic* or *just* vegetative.

7.3 Neuroimaging Brain Responses: Automaticity, Maps, and Interpretation

Although the compatibility of cortical responsiveness to sensory stimulation with VS is no longer controversial, interpretation of the meaning of the observed activations, even where they match what is typically seen in healthy volunteers under similar experimental conditions, is not always straightforward. First, analogously to behavioral (i.e., clinical) assessments of responsiveness, drawing the line between automatic and voluntary responses, is, at present, a controversial and not necessarily agreed upon task. Indeed, it is well established that many responses, behavioral or neuroimaging, fall into the category of automatic [17] and should thus not be taken to index anything else other than the presence of relatively intact neural circuits. In this respect, observing a brain response, in visual cortex, to an image is not much different from observing the contraction of the quadriceps in response to tapping the patellar ligament: both represent automatic response mechanisms. Second, assessing the presence of a cognitive process on the sole basis of brain response is an inductive (i.e., probabilistic), rather than deductive (i.e., certain), inference [18, 19]. In mathematical terms, the relationship tying the set of cognitive states that a person can enter and the set of brain activations that can be observed with neuroimaging is a non-injective function. That is to say, the mapping between the set of cognitive states and activation maps is not one-to-one. Rather, because of the relatively low spatial resolution of neuroimaging techniques, compared to the neural scale, multiple cognitive states can map onto the same pattern of brain activations, as detected with current BOLD fMRI, making the function tying the two sets non-invertible.

As illustrated in Fig. 7.1, while it is always possible to go from a given cognitive state (i.e., A, B, C, D, E) to specific brain activation maps (i.e., W, X, Y, Z), the reverse is not true. Given, for example, an observed pattern of fMRI brain activation X, it is not possible to say with certainty whether the generating cognitive state was

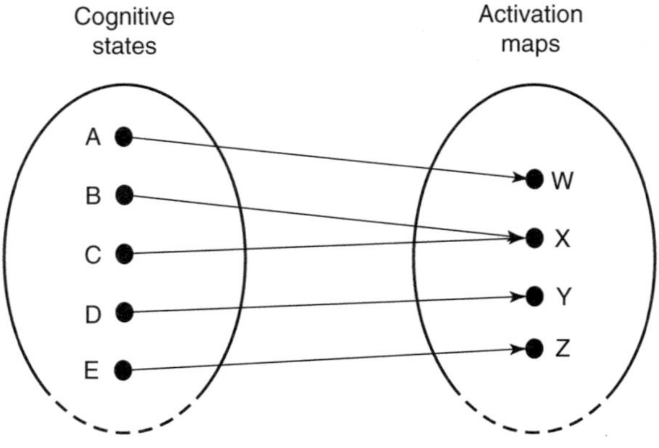

Fig. 7.1 Diagrammatic example of the non-injective nature of the relation between cognitive states and neuroimaging (fMRI) activation maps – given the current spatial resolution of the technique

B or C. In the context of disorders of consciousness, where a patient cannot report what cognitive activity she/he was engaging in, this means that observing a certain pattern of brain activation does not, by itself, allow one to unambiguously infer which specific cognitive processes the patient was engaged in. For this reason, it is conventional in neuroimaging studies to benchmark activation patterns detected in patients to those observed in healthy volunteers; to employ previous literature as a prior, suggesting which activations should be expected in patients; or, more rarely, to use convergent evidence from alternative approaches (e.g., anesthesia [20]). Finally, the relationship between brain activation and phenomenological *sensation* is the subject of intense research and debate and, in the context of severe brain injury, the source of some of the most complicated and controversial issues pertaining to the use of neuroimaging. Do activations in the so-called pain matrix, elicited by noxious stimulation, correspond to the *perception* of pain [14, 21]? Do brain activations in response to linguistic stimuli index comprehension [11, 20]? As discussed in the next section, the specific experimental design employed to elicit a given neural response is believed to be one of the determining factors in deciding how to interpret neuroimaging responses [22, 23].

7.4 Cogito Ergo Sum by fMRI[1]

Determining the degree of residual cognitive processing that may be available in patients who survive severe BI is very important and can assist, in the rehabilitative context, in determining which modalities might be – at least potentially – available to try to elicit responses in a patient and, where voluntary responses are detected, to harness them into methods of communication. An even more pressing – and more controversial – question, however, is whether neuroimaging responses can be taken to weigh on the determination of whether a patient is (at least minimally) conscious. In the insightful words of A.H. Ropper [24], is "*cogito ergo sum* by fMRI" possible (and/or admissible in the clinical context; see Chap. 12)? While the answer to this question is tied to several of the complexities mentioned above, many agree that one of the most important variables in adjudicating the issue of whether neuroimaging activations can index the presence of awareness is the nature of the experimental design employed (see [23, 25]). On the one hand, detecting different patterns of brain activation in response to *different* sensory stimulations, as in the picture of a face versus the picture of a house (see Fig. 7.2, top row), might be taken to imply that a brain possesses sufficient bottom-up mechanisms to distinguish the two stimuli (with no implication as to whether the patient has any subjective experience relating to them).

On the other hand, detecting different patterns of brain activation in response to the *same* stimulus, when under different instructions (i.e., mental sets), can only be explained – once sources of artifactual activation such as motion are excluded – by someone voluntarily complying with the instructions and engaging in top-down

[1] Title from [24]

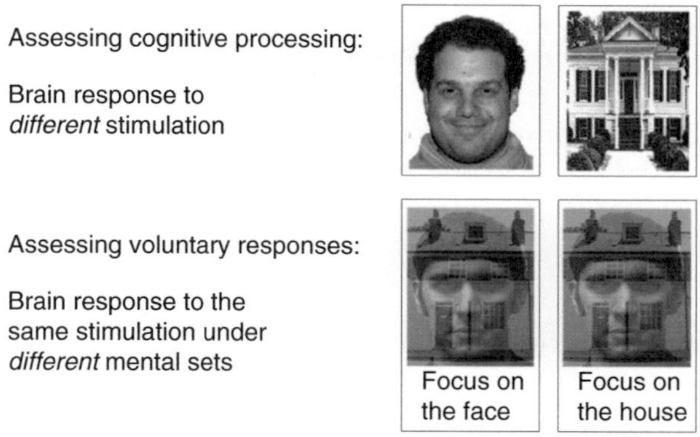

Fig. 7.2 *Top row*: Comparison of brain response to *different* sensory stimulations only allows inferring, generally speaking, "appropriate" bottom-up residual processing. Bottom *row:* Comparison of brain responses to the *same* sensory stimulation, but under *different* mental sets, allows inferring the presence of voluntary (i.e., top-down) responses (Adapted from [7])

voluntary cognitive processes, both of which imply a state of (at least minimal) awareness. To exemplify, consider the ambiguous stimulus depicted in the bottom row of Fig. 7.2. If a patient consistently demonstrates, in response to the same ambiguous picture, sustained (i.e., 16 s) upregulation of the face selective area in the fusiform gyrus and contemporaneous downregulation of the place-sensitive parahippocampal area in the periods in which she/he is asked to focus on the face in the image, and the reverse pattern in the periods in which she/he is asked to focus on the house in the image,[2] the simplest explanation is that the patient has understood the instructions and is performing the expected task.

In a landmark paper, in 2006, this logic was employed to demonstrate, for the very first time, that it is possible for a patient to appear unresponsive during bedside clinical assessments while being able to engage in voluntary top-down mental activities (e.g., "imagining playing tennis"), as captured by fMRI [26]. Specifically, a patient exhibited appropriate and sustained (for 30 s) brain activations, matching those seen in healthy volunteers, in response to short single-word cues (e.g., "tennis") when, by all clinical criteria, no sign of voluntary responsiveness or awareness could be detected. Since this pioneering paper, a number of publications have replicated the result in the context of similar experimental designs [27, 28], different experimental designs [29, 30], as well as different methodologies [31, 32].[3] While groundbreaking at the time, this finding is not necessarily unexpected. A number of previous studies, including retrospective audits [33, 34] as well as comparative evaluation of the diagnostic accuracy of different patient assessment techniques [35, 36], had already shown that misdiagnosis – by which (minimally) conscious patients are diagnosed as vegetative – is more frequent than desirable. While the relative rarity of these conditions, inconsistent terminology, and lack of specialized training were recognized early on as important

[2] Which is the pattern of activation observed in healthy volunteers performing the same task

[3] Even if, currently, the interpretation of the results presented in [31] has been subject to debate

causes of misdiagnosis [34], it is now also understood that the diagnosis of VS rests on a logical flaw [22]: absence of observable evidence of awareness (i.e., failure of a patient to demonstrate any recognizably voluntary behavior) cannot be taken to necessarily imply absence of awareness. For, a patient could be (minimally) conscious, but unable to respond because of sensory or motor impairments or transient unconsciousness, or even just unwillingness to respond [4].

7.5 Important Caveats

As the potential of neuroimaging to uncover signs of awareness in otherwise behaviorally unresponsive patients is more and more recognized, it is important to be mindful of a number of important issues concerning the interpretation of neuroimaging data in this context.

7.6 Positives and Negatives: Dissociation Is a 2-Way Street

As briefly reviewed above, a number of studies have shown that it is possible for a patient to appear unresponsive in (behavior-based) bedside clinical testing while being responsive in neuroimaging assessments [26, 27, 29, 30, 32]. A number of recent studies, however, have reported instances of the reverse dissociation: some patients can demonstrate a state of consciousness at the bedside but fail to show any significant activation during "active" neuroimaging sessions. Bardin and colleagues, for example, have shown that only half of a group of (at least minimally) conscious patients, as determined by clinical testing, could demonstrate significant activity in mental imagery tasks [37]. Similarly, in a recent study of top-down processes in DOC patients, while 3 out of 8 patients with a VS diagnosis could demonstrate voluntary (brain) behavior during neuroimaging sessions, only 6 out of 12 MCS patients[4] and 3 out of 4 exit-MCS patients could not demonstrate any significant activation [29]. In a large cohort study, the overall sensitivity of the "imagery" task in fMRI to detect a state of MCS was recently estimated at about 45 % [28]. The existence of two-directional dissociations highlights two important aspects of the use of neuroimaging in the context of DOC. On the one hand, it confirms that there are instances in which neuroimaging can uncover voluntary brain responses in a subset of patients who appear behaviorally (i.e., at the bedside) unresponsive. On the other hand, these dissociations imply that false negatives are as possible in neuroimaging assessments as they are in behavioral assessments (although it is currently unknown how the two rates compare). In other words, negative neuroimaging results should not be interpreted.

[4] An additional 4 MCS patients were included in the analysis, but because of excessive movement, their data could not be meaningfully analyzed – another important issue in the domain of neuroimaging assessments.

7.7 The "Tip of the Iceberg" Problem

An important, but not very appreciated, issue, in the use of neuroimaging to assess DOC patients, is the "tip of the iceberg" phenomenon. The kind of "active" tasks (e.g., mental imagery [26], target monitoring [38]) currently employed to covertly detect the presence of consciousness with neuroimaging requires the concurrent presence of a number of cognitive processes in addition to a state of awareness. To name a few, patients must, at a minimum, retain language capabilities sufficient to comprehend a set of instructions; memory functions sufficient to allow maintaining a set of instructions throughout an experimental run; sensory resources, in all the relevant modalities, sufficient to allow processing stimuli; as well as executive functions sufficient to allow, for example, periodic engaging and disengaging in the relevant mental task. Although many of these problems are common to standard clinical assessments, they further stress the importance of not interpreting negative results as evidence of unconsciousness, as well as the need to develop non-language-based and, ideally, "passive" neuroimaging tests capable of detecting neural markers of a conscious state (e.g., [15, 39, 40]).

7.8 Unconscious or in a "Living Hell": A (Probably) False Dichotomy

Finally, a last important *caveat* applies to the over-interpretation of evidence (neuroimaging or otherwise) in the context of DOC. In the 1990 ruling of the Supreme Court of the United States (497 US 261) in the case of Cruzan v. Director, Missouri Department of Health, Judge Blackmar, after pointing out that the patient in question, Nancy Cruzan, might have exhibited (probably reflexive) responses to noxious stimulation, noted that "If she has any awareness of her surroundings, her life must be a living hell." While we will not visit the important legal and ethical issues surrounding DOC (see Chap. 14), there is an important neuroscientific consideration to make with respect to how much mental life should be attributed, in the absence of direct evidence, to patients demonstrating neuroimaging (or for that matter, behavioral) responses. Often, because most fMRI experiments benchmark the activations seen in patients to those observed in healthy volunteers (e.g., [7, 26, 29, 37]), it is tempting to infer that, where matching activations are seen, these imply that the patient might possess the same state of awareness of healthy individuals. While it is certainly not impossible for a behaviorally unresponsive patient to retain normal consciousness (as would be in the very different condition of complete locked-in syndrome), it must be recognized that the brain of an MCS patients, even if capable of supporting some level of consciousness, is nonetheless structurally and functionally severely pathological and extremely different from that of a healthy individual. PET data, for example, have clearly shown that the MCS brain is severely hypometabolic, presenting a cerebral metabolic rate of glucose at about 55 % that

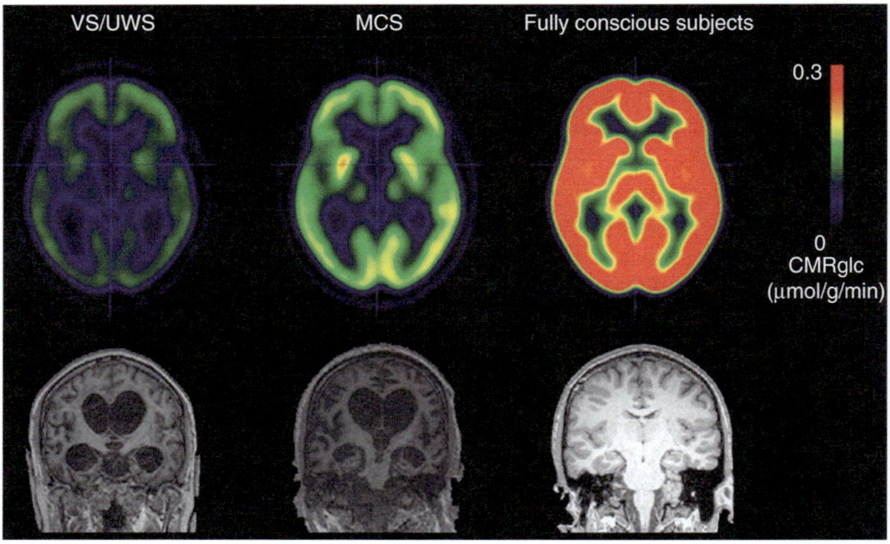

Fig. 7.3 *Top*: Depiction of average metabolic rates observed in VS, MCS, and healthy volunteers (Adapted from [42] under Creative Commons CC-BY license). *Bottom*: Sample T1-weighted structural images for a VS, MCS, and healthy volunteer, viewed in a coronal cut (note: the brains shown in the *bottom row* do not correspond to the brains in the *top row* and are only meant to represent a sample individual from each group)

observed in healthy volunteers. In fact, although local differences between conscious and unconscious patients can be detected, MCS metabolic rate is much closer to that of a VS patients (estimated at approximately 42 % that of the healthy brain) than fully conscious, healthy, individuals [41, 42] (see Fig. 7.3).

Similarly, as illustrated in the bottom row of Fig. 7.3, despite being (at least minimally) conscious, MCS patients present severe widespread pathology as compared to healthy volunteers, which has been measured in terms of both subcortical atrophy [43] as well as degradation of cortico-cortical and cortico-subcortical connectivity [44]. Thus, observing in a DOC patient, voluntary brain responses matching those seen in healthy volunteers should not be automatically taken, by itself, to imply an equivalence between the state of consciousness of the patient and that of a healthy individual – which should, however, not be excluded a priori [45]. Rather, the full power of neuroimaging techniques could be employed, in otherwise nonresponsive patients, to address in an evidence-based way the question of which cognitive resources a patient might retain [46].

7.9 Conclusion

In all, regardless of where one falls on the many issues raised above, there is little doubt that the use of in vivo neuroimaging has greatly benefited the field of disorders of consciousness. First, in the past 18 years, these techniques have started revealing the (at first surprising) extent to which cortical processing can be retained in VS as well as MCS patients despite having sustained a catastrophic brain injury (e.g., [7–9, 11–14, 47–50]). These results fully highlight the ramifications, in the clinical context, of our limited scientific understanding of what consciousness is and our inability to directly measure it or quantify it objectively [22, 23]. Furthermore, neuroimaging has also shown that it is possible for a patient to appear, clinically (i.e., behaviorally), unresponsive while being in fact (at least) minimally conscious, as indexed by the ability to voluntarily engage and disengage, in response to verbal commands, in top-down mental tasks (e.g., [26, 27, 29, 30, 32]). It is crucial to stress that these (to date relatively few) cases in which a dissociation has been reported between the level of awareness observable at the bedside and that observable in neuroimaging assessments do not only reflect the natural challenges tied to assessing patients that have sustained severe brain injuries [33–36]. Indeed, as discussed above and elsewhere [4, 23, 25], no matter how skilled an assessor, a (minimally) conscious patient unable to manifest, through muscle-dependent responses, her state of awareness – due, for example, to motor impairment – would be impossible to distinguish from a VS patient on the basis of clinical protocols. It is in these cases that the full value of neuroimaging becomes evident.

Still, it is also undeniable that the use of neuroimaging techniques in the context of disorders of consciousness requires careful consideration of a number of issues. First, the interpretation of brain activations is necessarily secondary to the specific experimental paradigm employed [25]. Only under certain experimental circumstances (unless convergent evidence from other methodologies, such as anesthesia, is available; see [20]) can brain responses be taken to mark the presence of a state of minimal awareness. Second, only positive evidence should be interpreted, because negative findings are unable to differentiate cases of truly VS patients from cases of MCS patients unable (or unwilling) to respond. Finally, while DOC patients exhibit highly pathological brain function and structure [42, 43, 51], it is also true that the human brain can maintain surprisingly high level of functioning even in the presence of severely pathological features (e.g., [45]; see also the degree of pathology evident in some "responsive" DOC patients [27, 29, 30, 37]). Thus, wherever behavior-based evidence cannot be obtained, neuroimaging might be the one approach capable of assessing which, and how many, cognitive processes can be imputed to any given DOC patient [46].

References

1. Monti MM, Laureys S, Owen AM (2010) The vegetative state. BMJ 341:c3765. doi:10.1136/bmj.c3765
2. Fins JJ, Schiff ND (2006) Shades of gray: new insights into the vegetative state. Hastings Cent Rep 36(6):8
3. Laureys S (2005) The neural correlate of (un)awareness: lessons from the vegetative state. Trends Cogn Sci 9(12):556–559. doi:10.1016/j.tics.2005.10.010
4. Monti MM (2012) Cognition in the vegetative state. Annu Rev Clin Psychol 8:431–454. doi:10.1146/annurev-clinpsy-032511-143050
5. Jennett B, Plum F (1972) Persistent vegetative state after brain damage. RN 35(10):ICU1–ICU4

6. Menon DK, Owen AM, Williams EJ, Minhas PS, Allen CM, Boniface SJ et al (1998) Cortical processing in persistent vegetative state Wolfson Brain Imaging Centre Team. Lancet 352(9123):200
7. Monti MM, Pickard JD, Owen AM (2013) Visual cognition in disorders of consciousness: from V1 to top-down attention. Hum Brain Mapp 34(6):1245–1253. doi:10.1002/hbm.21507
8. Owen AM, Menon DK, Johnsrude IS, Bor D, Scott SK, Manly T et al (2002) Detecting residual cognitive function in persistent vegetative state. Neurocase 8(5):394–403
9. Boly M, Faymonville ME, Peigneux P, Lambermont B, Damas P, Del Fiore G et al (2004) Auditory processing in severely brain injured patients: differences between the minimally conscious state and the persistent vegetative state. Arch Neurol 61(2):233–238. doi:10.1001/archneur.61.2.233
10. Laureys S, Faymonville ME, Degueldre C, Fiore GD, Damas P, Lambermont B et al (2000) Auditory processing in the vegetative state. Brain 123(Pt 8):1589–1601
11. Coleman MR, Davis MH, Rodd JM, Robson T, Ali A, Owen AM et al (2009) Towards the routine use of brain imaging to aid the clinical diagnosis of disorders of consciousness. Brain 132(Pt 9):2541–2552. doi:10.1093/brain/awp183
12. Coleman MR, Rodd JM, Davis MH, Johnsrude IS, Menon DK, Pickard JD et al (2007) Do vegetative patients retain aspects of language comprehension? evidence from fMRI. Brain 130(Pt 10):2494–2507
13. Laureys S, Faymonville ME, Peigneux P, Damas P, Lambermont B, Del Fiore G et al (2002) Cortical processing of noxious somatosensory stimuli in the persistent vegetative state. Neuroimage 17(2):732–741
14. Boly M, Faymonville ME, Schnakers C, Peigneux P, Lambermont B, Phillips C et al (2008) Perception of pain in the minimally conscious state with PET activation: an observational study. Lancet Neurol 7(11):1013–1020. doi:10.1016/S1474-4422(08)70219-9
15. Bekinschtein TA, Dehaene S, Rohaut B, Tadel F, Cohen L, Naccache L (2009) Neural signature of the conscious processing of auditory regularities. Proc Natl Acad Sci U S A 106(5):1672–1677. doi:10.1073/pnas.0809667106
16. Kotchoubey B, Jetter U, Lang S, Semmler A, Mezger G, Schmalohr D et al (2006) Evidence of cortical learning in vegetative state. J Neurol 253(10):1374–1376. doi:10.1007/s00415-006-0221-0
17. Dehaene S, Naccache L, Le Clec HG, Koechlin E, Mueller M, Dehaene-Lambertz G et al (1998) Imaging unconscious semantic priming. Nature 395(6702):597–600
18. Henson R (2005) What can functional neuroimaging tell the experimental psychologist? Q J Exp Psychol A 58(2):193–233. doi:10.1080/02724980443000502
19. Poldrack RA (2006) Can cognitive processes be inferred from neuroimaging data? Trends Cogn Sci 10(2):59–63. doi:10.1016/j.tics.2005.12.004
20. Davis MH, Coleman MR, Absalom AR, Rodd JM, Johnsrude IS, Matta BF et al (2007) Dissociating speech perception and comprehension at reduced levels of awareness. Proc Natl Acad Sci U S A 104(41):16032–16037. doi:10.1073/pnas.0701309104
21. Schnakers C, Zasler ND (2007) Pain assessment and management in disorders of consciousness. Curr Opin Neurol 20(6):620–626. doi:10.1097/WCO.0b013e3282f169d9
22. Monti MM, Coleman MR, Owen AM (2009) Neuroimaging and the vegetative state: resolving the behavioralassessmentdilemma?AnnNYAcadSci1157:81–89.doi:10.1111/j.1749-6632.2008.04121.x
23. Owen AM, Coleman MR (2008) Functional neuroimaging of the vegetative state. Nat Rev Neurosci 9(3):235–243. doi:10.1038/nrn2330
24. Ropper AH (2010) Cogito ergo sum by MRI. N Engl J Med 362(7):648–649. doi:10.1056/NEJMe0909667
25. Monti MM, Owen AM (2010) Behavior in the brain using functional neuroimaging to assess residual cognition and awareness after severe brain injury. J Psychophysiol 24(2):76–82. doi:10.1027/0269-8803/A000016
26. Owen AM, Coleman MR, Boly M, Davis MH, Laureys S, Pickard JD (2006) Detecting awareness in the vegetative state. Science 313(5792):1402
27. Monti MM, Vanhaudenhuyse A, Coleman MR, Boly M, Pickard JD, Tshibanda L et al (2010) Willful modulation of brain activity in disorders of consciousness. N Engl J Med 362(7):579–589. doi:10.1056/NEJMoa0905370

28. Stender J, Gosseries O, Bruno MA, Charland-Verville V, Vanhaudenhuyse A, Demertzi A et al (2014) Diagnostic precision of PET imaging and functional MRI in disorders of consciousness: a clinical validation study. Lancet 384(9942):514–522. doi:10.1016/S0140-6736(14)60042-8
29. Monti MM, Rosenberg M, Finoia P, Kamau E, Pickard JD, Owen AM (2015) Thalamo-frontal connectivity mediates top-down cognitive functions in disorders of consciousness. Neurology 84(2):167–173
30. Naci L, Owen AM (2013) Making every word count for nonresponsive patients. JAMA Neurol 70(10):1235–1241. doi:10.1001/jamaneurol.2013.3686
31. Cruse D, Chennu S, Chatelle C, Bekinschtein TA, Fernandez-Espejo D, Pickard JD et al (2011) Bedside detection of awareness in the vegetative state: a cohort study. Lancet 378(9809): 2088–2094. doi:10.1016/S0140-6736(11)61224-5
32. John ER, Halper JP, Lowe RS, Merkin H, Defina P, Prichep LS (2011) Source imaging of QEEG as a method to detect awareness in a person in vegetative state. Brain Inj 25(4): 426–432. doi:10.3109/02699052.2011.558045
33. Andrews K, Murphy L, Munday R, Littlewood C (1996) Misdiagnosis of the vegetative state: retrospective study in a rehabilitation unit. BMJ 313(7048):13–16
34. Childs NL, Mercer WN, Childs HW (1993) Accuracy of diagnosis of persistent vegetative state. Neurology 43(8):1465–1467
35. Schnakers C, Giacino J, Kalmar K, Piret S, Lopez E, Boly M et al (2006) Does the four score correctly diagnose the vegetative and minimally conscious states? Ann Neurol 60(6):744–745, author reply 5
36. Schnakers C, Vanhaudenhuyse A, Giacino J, Ventura M, Boly M, Majerus S et al (2009) Diagnostic accuracy of the vegetative and minimally conscious state: clinical consensus versus standardized neurobehavioral assessment. BMC Neurol 9:35. doi:10.1186/1471-2377-9-35
37. Bardin JC, Fins JJ, Katz DI, Hersh J, Heier LA, Tabelow K et al (2011) Dissociations between behavioural and functional magnetic resonance imaging-based evaluations of cognitive function after brain injury. Brain 134(Pt 3):769–782. doi:10.1093/brain/awr005
38. Monti MM, Coleman MR, Owen AM (2009) Executive functions in the absence of behavior: functional imaging of the minimally conscious state. Prog Brain Res 177:249–260. doi:10.1016/S0079-6123(09)17717-8
39. Chennu S, Finoia P, Kamau E, Allanson J, Williams GB, Monti MM et al (2014) Spectral signatures of reorganised brain networks in disorders of consciousness. PLoS Comput Biol 10(10):e1003887. doi:10.1371/journal.pcbi.1003887
40. Monti MM, Lutkenhoff ES, Rubinov M, Boveroux P, Vanhaudenhuyse A, Gosseries O et al (2013) Dynamic change of global and local information processing in propofol-induced loss and recovery of consciousness. PLoS Comput Biol 9(10):e1003271. doi:10.1371/journal.pcbi.1003271
41. Laureys S, Owen AM, Schiff ND (2004) Brain function in coma, vegetative state, and related disorders. Lancet Neurol 3(9):537–546. doi:10.1016/S1474-4422(04)00852-X
42. Stender J, Kupers R, Rodell A, Thibaut A, Chatelle C, Bruno M-A et al (2015) Quantitative rates of brain glucose metabolism distinguish minimally conscious from vegetative state patients. J Cereb Blood Flow Metab 35(1):58–65. doi:10.1038/jcbfm.2014.169
43. Lutkenhoff ES, Chiang J, Tshibanda L, Kamau E, Kirsch M, Pickard JD et al (2015) Thalamic and extrathalamic mechanisms of consciousness after severe brain injury. Ann Neurol 78(1):68–76
44. Fernandez-Espejo D, Soddu A, Cruse D, Palacios EM, Junque C, Vanhaudenhuyse A, Fernandez-Espejo D, Soddu A, Cruse D, Palacios EM, Junque C, Vanhaudenhuyse A et al (2012) A role for the default mode network in the bases of disorders of consciousness. Ann Neurol 72(3):335–343. doi:10.1002/Ana.23635
45. Feuillet L, Dufour H, Pelletier J (2007) Brain of a white-collar worker. Lancet 370(9583):262. doi:10.1016/S0140-6736(07)61127-1
46. Peterson A, Naci L, Weijer C, Cruse D, Fernández-Espejo D, Graham M et al (2013) Assessing decision-making capacity in the behaviorally nonresponsive patient with residual covert awareness. AJOB Neuroscience 4(4):3–14

47. Chennu S, Finoia P, Kamau E, Monti MM, Allanson J, Pickard JD et al (2013) Dissociable endogenous and exogenous attention in disorders of consciousness. Neuroimage Clin 3:450–461. doi:10.1016/j.nicl.2013.10.008
48. Di HB, Yu SM, Weng XC, Laureys S, Yu D, Li JQ et al (2007) Cerebral response to patient's own name in the vegetative and minimally conscious states. Neurology 68(12):895–899
49. Giacino JT, Hirsch J, Schiff N, Laureys S (2006) Functional neuroimaging applications for assessment and rehabilitation planning in patients with disorders of consciousness. Arch Phys Med Rehabil 87(12 Suppl 2):S67–S76
50. Owen AM, Coleman MR, Menon DK, Johnsrude IS, Rodd JM, Davis MH et al (2005) Residual auditory function in persistent vegetative state: a combined PET and fMRI study. Neuropsychol Rehabil 15(3–4):290–306
51. Fernandez-Espejo D, Bekinschtein T, Monti MM, Pickard JD, Junque C, Coleman MR et al (2011) Diffusion weighted imaging distinguishes the vegetative state from the minimally conscious state. Neuroimage 54(1):103–112. doi:10.1016/j.neuroimage.2010.08.035

Chapter 8
Exploring the Neurophysiological Correlates of Loss and Recovery of Consciousness: Perturbational Complexity

Silvia Casarotto, Mario Rosanova, Olivia Gosseries, Mélanie Boly, Marcello Massimini, and Simone Sarasso

Abstract Although assessing a subject's capacity for consciousness is commonly based on an input-output paradigm, clinical and physiological evidence advocate the development of brain-based metrics that are independent on both sensory inputs and motor outputs. As an attempt in this direction, we devised an empirical measure of complexity derived from the electroencephalographic (EEG) response to a direct cortical perturbation with transcranial magnetic stimulation (TMS), the perturbational complexity index (PCI). Based on theoretical postulates, PCI gauges the conjoint presence of integration and information in the human brain, independently of sensory inputs and motor outputs. In a preliminary series of experiments, PCI was tested on TMS-evoked potentials recorded in healthy subjects during wakefulness, dreaming, NREM sleep, and different levels of sedation induced by different anesthetic agents, as well as in patients who emerged from coma and attained a stable diagnosis. These experiments show that PCI allows a reliable assessment of the level of consciousness at the single-subject level and prompt further validation toward the development of a diagnostic test. In parallel, elucidating the mechanisms by which brain complexity collapses and recovers following brain injury may provide novel insight on the physiopathology and the treatment of disorders of consciousness.

Keywords Consciousness • Neurophysiology • Perturbational complexity index • TMS

S. Casarotto • M. Rosanova • M. Massimini • S. Sarasso (✉)
Department of Biomedical and Clinical Sciences Luigi Sacco,
University of Milan, Milan, Italy
e-mail: simone.sarasso@unimi.it

O. Gosseries • M. Boly
Coma Science Group, Cyclotron Research Center and Neurology Department,
University and University Hospital of Liège, Liège, Belgium

M.M. Monti, W.G. Sannita (eds.), *Brain Function and Responsiveness in Disorders of Consciousness*, DOI 10.1007/978-3-319-21425-2_8

8.1 The Problem of Disconnected Consciousness

In everyday life and in the clinical practice, assessing consciousness relies ultimately on a behavioral input-output paradigm; we stimulate subjects in different sensory modalities and we observe their motor responses. If an individual is able to connect to the surrounding environment and then produces appropriate responses, she/he is considered conscious. Thanks to recent conceptual and technical advances, a similar input-output paradigm can now be applied even to patients who are fully paralyzed and cannot engage in any behavioral output, such as patients affected by a total locked-in syndrome [1–3]. In this case, subjects are still presented with sensory inputs, but the blockage of motor output is bypassed by recording directly neuronal activations by means of functional neuroimaging (fMRI) and electroencephalography (EEG). In this way, it is possible to detect specific neuronal markers of mental imagery being covertly performed in response to sensory stimuli and commands, thus demonstrating that some patients are aware despite their lack of behavioral responsiveness. These neuroimaging paradigms leave no doubts in case of a positive result: if the neuronal output is consistent and specific with respect to the sensory input, the subject is conscious. However, the neuroimaging approach lacks sensitivity and many conscious subjects fail to produce the appropriate neuronal responses to stimulations and commands [4].

A fundamental problem is that conscious experience can also be present in subjects who are disconnected from the environment on the input side. For example, during dreaming complex, temporally unfolding hallucinatory episodes can be as intense and vivid as waking consciousness—yet sensory stimuli are ignored to the point that they are rarely incorporated in the experience [5]. Consciousness may completely disconnect from the external environment also during some forms of anesthesia. Some dissociative anesthetic agents, such as ketamine at high doses, are known to induce a dreamlike hallucinatory state associated with sensory disconnection and complete unresponsiveness. Similar disconnections may occur in pathological conditions whereby a brain-injured subject may not respond to verbal commands or sensory stimuli because a peripheral or central lesion prevents sensory inputs from being transmitted and processed effectively. In all these cases, detecting consciousness based on an input-output paradigm may result in a significant rate of false negatives.

Clearly, finding a way to detect consciousness independent of both sensory inputs and motor outputs is a daunting proposition. One should first postulate which neural mechanisms are fundamental for consciousness, then should devise a practical way to gauge these mechanisms, and, finally, should validate this novel metric across several physiological, pharmacological, and pathological conditions in a large cohort of subjects who are able to provide an immediate (or retrospective) report about their level of consciousness. This validation should start from typical conditions in which consciousness is present (wakefulness) or lost (deep NREM sleep, seizures, anesthesia) and then extend to more ambiguous cases in which consciousness may be present but disconnected (dreaming, ketamine anesthesia). To the extent that this long process yields an index that is highly specific and highly sensitive for the presence of an immediate, or retrospective, conscious report across all these conditions, it may be possible to obtain a reliable frame of reference to detect a capacity for consciousness in patients who are completely disconnected from the external environment. Below, we describe a preliminary attempt in this direction.

8.2 Consciousness and Brain Complexity

The first step requires formulating a postulate. A parsimonious approach is to start from self-evident axioms in order to establish what physical properties are fundamental for consciousness and how they can be measured. Naturally, in the case of consciousness, evidence can only be gathered from phenomenology, the first-person observation of subjective experience itself [6]. Phenomenologically, each conscious experience is both differentiated—that is, it has many specific features that distinguish it from a large repertoire of other experiences—and integrated—that is, it cannot be divided into independent components. Neurophysiologically, these fundamental properties of subjective experience rely on the ability of multiple, functionally specialized modules of the thalamocortical system to interact rapidly and effectively to form an integrated whole [7–10]. Hence, the fundamental postulate is that consciousness requires an optimal balance between functional integration and functional differentiation in thalamocortical networks—otherwise defined as brain complexity [11–14]. This notion implies that the complexity of brain activity should be high when consciousness is present and low whenever consciousness is lost in sleep, anesthesia, or coma [12, 15, 16].

8.3 Assessing Brain Complexity Through Perturbations

The second problem entails devising an empirical measurement that must be, at the same time, practical and in line with the fundamental postulate. On the one hand, theoretical measures that have been designed to assess the joint presence of differentiation and integration in neural systems [17–19] are only applicable to simple systems of simulated elements or under highly restrictive assumptions and have not been tested on human brains. On the other hand, currently used empirical indices of consciousness are either based on integration alone [as judged by the extent or synchronization of cortical activations [20–22]] or on differentiation alone [as judged by entropy or spectral content [23, 24]] and do not reliably assess consciousness in individual patients or across many different conditions.

A viable and principled way to gauge the conjoint presence of integration and information in real brains involves directly probing the cerebral cortex (in order to avoid possible subcortical filtering and gating) by employing a perturbational approach (thus testing causal interactions rather than temporal correlations) and examining to what extent cortical regions can interact as a whole (integration) to produce differentiated responses (information) [25]. Practically, this approach can be applied to the human brain by employing a combination of transcranial magnetic stimulation (TMS) and high-density EEG, a technique that allows stimulating directly a subset of cortical neurons and measuring, with good spatial-temporal resolution, the effects produced by this perturbation on the rest of the thalamocortical system [26]. According to this proposal, a signature of consciousness is that the thalamocortical system should respond to TMS with complex, rapidly changing activity patterns (information) that affect a distributed set of cortical areas (integration). On the other hand, it can be predicted that during loss of consciousness, whether this is caused by sleep, anesthesia, or coma, the brain should react to perturbations with a response that is local (loss of integration) and/or stereotypical (loss of information) (Fig. 8.1a). In order to quantify the spatiotemporal complexity of TMS-evoked cortical activations, a novel empirical measure called the perturbational complexity index (PCI) has been recently introduced [27]. Calculating PCI involves two fundamental steps: (1) perturbing the cortex with TMS to engage distributed interactions in the brain (*integration*) and (2) "zipping" (i.e., compressing) the resulting electrocortical responses to measure their algorithmic complexity (*information*) (Fig. 8.1b). The underlying idea is that PCI should be low if causal interaction among cortical areas is reduced (loss of integration), because the matrix of activation engaged by TMS is spatially restricted; PCI is also expected to be low if many interacting areas react to the perturbation but they do so in a stereotypical way (loss of differentiation) because, in this case, the resulting matrix is large but redundant and can be effectively compressed. In fact, PCI should reach high values only if the initial perturbation is transmitted to a large set of integrated areas that react in a differentiated way, giving rise to a spatiotemporal pattern of deterministic activation that cannot be easily reduced.

8.4 Testing the Perturbational Complexity Index in Sleep Anesthesia and Coma

In a preliminary series of experiments, PCI was tested on a large dataset of TMS-evoked potentials recorded in healthy subjects during wakefulness, dreaming, NREM sleep, and different levels of sedation induced by different anesthetic agents (midazolam, xenon, and propofol), as well as in patients who emerged from coma and attained a stable diagnosis as assessed through repeated measurements with the Coma Recovery Scale-Revised [28] (overall, 208 sessions in 52 subjects). In healthy, awake subjects, TMS triggered a complex EEG response involving different cortical areas at different times (Fig. 8.1c) [29, 30]. When subjects lost consciousness during NREM sleep, TMS pulses invariably produced a simple wave of activation that remained localized to the site of stimulation, indicating a breakdown of communication and a loss of integration within thalamocortical networks (Fig. 8.1c) [29].

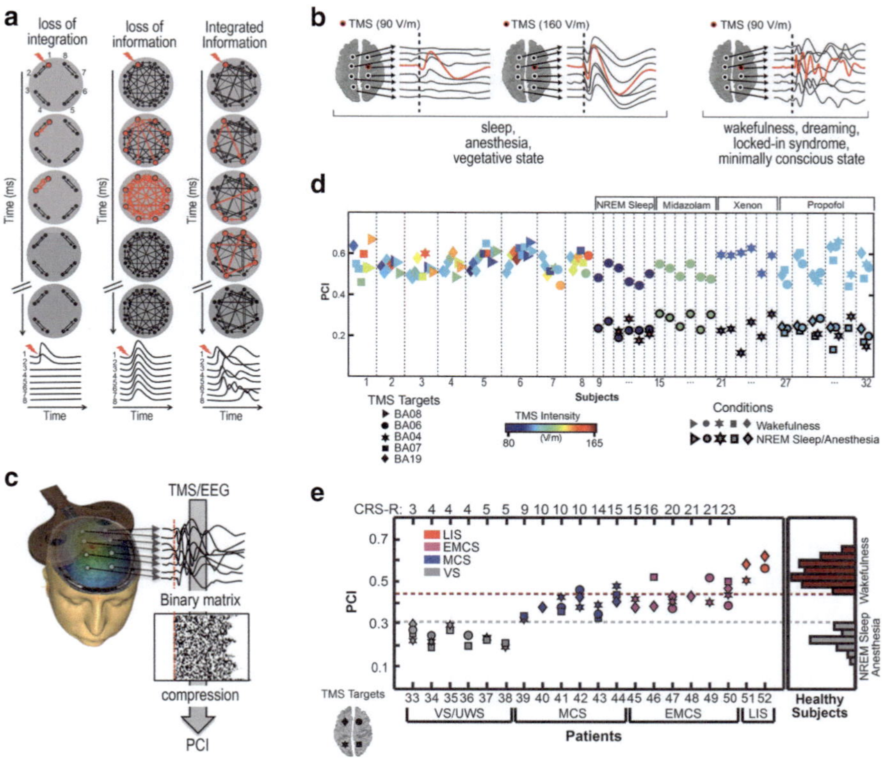

Fig. 8.1 TMS/EEG allows to test the IIT in humans and PCI quantifies the spatiotemporal pattern of deterministic brain response to perturbation. (**a**) Outline of the possible responses to perturbation of an idealized system. When the system is not integrated (*left*), the response is short-leaved and limited to locally connected elements; when the system is fully connected in a uniform way (*middle*), the response spreads in all the elements at the same time and then rapidly vanishes; when the system realizes an optimal balance between information and integration, the response is long-lasting and widespread, yet complex, involving different elements at different times (*right*). (**b**) Examples of the typical TMS-evoked EEG responses obtained when consciousness is lost and present. (**c**) PCI is computed by compressing the binary matrix of significant spatiotemporal activations evoked by TMS at the source level estimated from scalp recordings. (**d**, **e**) PCI distributions in healthy subjects during wakefulness and conditions of loss of consciousness and in brain-injured vegetative state, minimally conscious state and emergence from MCS patients

The disappearance of a long-range, differentiated pattern of cortical activation was not simply due to a reduction of responsiveness of hyperpolarized cortical neurons. In fact, increasing TMS intensity only resulted in a larger, simple positive-negative wave, closely resembling a spontaneous sleep slow wave [31]. In this case, the response to TMS was stereotypical and spread like an oil spot to vast regions of the cortex, revealing a loss of differentiation (Fig. 8.1c). Similar local and/or stereotypical responses were invariably also found during general anesthesia [32] as well as in brain-injured patients with an unambiguous clinical diagnosis of a vegetative state [33] (Fig. 8.1c). Crucially, wakefulness-like, complex responses always recovered during REM sleep [34] in minimally conscious patients, who showed signs of non-reflexive activity, and in locked-in syndrome (LIS) subjects, who were totally paralyzed except for vertical eye movements through which they signal that they were aware [33].

Accordingly, in healthy (awake, sleeping, and anesthetized) subjects—whose level of consciousness can be known based on subjective reports upon awakening—PCI provided a reproducible scale along the unconsciousness-consciousness spectrum (Fig. 8.1d). Notably, brain-injured patients (MCS, EMCS, and LIS) who showed minimal signs of consciousness attained values of brain complexity that were invariably above the maximum value obtained in unconscious, anesthetized, or sleeping, healthy subjects (Fig. 8.1e).

8.5 Toward the Development of a Reliable Diagnostic Tool

The preliminary results obtained so far are encouraging but further testing will be needed before PCI can be practically employed at the bedside of disorders of consciousness (DOC) patients. First, the index needs to be validated in a larger cohort of healthy subjects, who can provide a report about their level of consciousness. This cohort should include conscious subjects affected by brain lesions with different extents and etiologies (stroke, traumatic injury, postanoxic lesions, late-stage Alzheimer disease). This testing will provide a thorough assessment of the sensitivity of PCI as well as indications on how to optimally perturb the brain with TMS in the presence of focal cortical damage or diffuse alterations. As already mentioned, PCI measurements should also be extended to conditions in which subjects are temporarily disconnected but report conscious experience retrospectively. This will involve experiments during dreaming (both in REM and NREM sleep) as well as during ketamine anesthesia. Overall, the values of PCI obtained in well-controlled conditions in which consciousness is known to be present (report, or delayed report) or absent (no report) will then be used to construct a receiver operator characteristic (ROC) analysis and to derive a cutoff that maximizes accuracy of distinction between these two conditions (Youden's index). It will then be crucial to verify whether this cutoff confirms the sensitivity of PCI in a large cohort of minimally conscious state (MCS) patients, as identified by the CRS-R. This is a key test, since current brain-derived measures of consciousness have a high rate of false negatives in this category of patients. At this point, a diagnostic test with an established accuracy will be available to assess consciousness independently of sensory inputs and motor outputs. This diagnostic measure could be finally applied to unresponsive wakefulness syndrome (UWS) patients, whose level of consciousness is unknown. For example, finding a PCI value well above the cutoff derived from the report/no report distributions in one of these patients would suggest that she/he is conscious to some extent, albeit completely disconnected from the external environment.

8.6 Exploring the Neurophysiological Mechanisms Underlying Loss and Recovery of Consciousness

Establishing a link between consciousness and brain complexity—in both theory and practice—may also help shedding new light on the cortical mechanisms of loss and recovery of consciousness in brain-injured patients. To this aim, once again, the contrast between wakefulness and sleep offers an interesting starting point. In fact, as previously discussed, TMS-evoked EEG responses obtained in stable chronic VS/UWS patients show a remarkable similarity with those obtained during NREM sleep, thus pointing to a common electrocortical event in both these conditions.

As already described above (Fig. 8.1c), during NREM, the differentiated pattern of deterministic interactions that propagate through a distributed network of cortical areas triggered by TMS during wakefulness is replaced by a simple response that remains either local [29] or spreads like an oil spot, often matching the EEG criteria for a sleep slow wave, or a K-complex [31]. These EEG graphoelements are typically occurring during natural sleep and are associated with brief periods of neuronal membrane hyperpolarization and silence (the so-called down state) [35] known to be due to neuronal bistability. Bistability, in turn, is caused by an increased activity of leak K+ channels brought about by decreased brainstem cholinergic activity and defined as the tendency of cortical neurons to fall into a silent down state in response to transient increases in activity [36, 37].

We hypothesized that bistability may drastically impair the emergence of complex patterns of causal interactions among cortical areas. Specifically, we hypothesize that (i) during NREM a group of neurons that receives a cortical input rapidly plunges into a down state and that (ii) this period of silence breaks off the causal effects of the initial input.

In a recent study [38], we tested this hypothesis by using intracranial single-pulse electrical stimulation (SPES) and simultaneous stereotactic EEG (SEEG) recordings in 8 patients undergoing neurosurgical evaluation for intractable epilepsy during both wakefulness and NREM sleep. Intracranial recordings allow a reliable, although indirect, detection of cortical down states in terms of significant suppression of high-frequency power above 20 Hz in the LFP [39, 40]. Also, intracranial perturbations with SPES allow assessing cortico-cortical and/or cortico-subcorticocortical interactions from a causal perspective [41].

We observed that, while during wakefulness electrical stimulation triggers a chain of deterministic phase-locked activations in its cortical targets, during NREM the same input induces a slow wave associated with an OFF period (suppression of power above 20 Hz), possibly reflecting a neuronal down state. Crucially, after the OFF period, cortical activity resumes to wakefulness-like levels, but the deterministic effects of the initial input are lost, as indicated by a sharp drop of phase-locked activity (Fig. 8.2). Overall, these results suggest that the intrinsic tendency of cortical neurons to fall into a down state after a transient activation (i.e., bistability) prevents the emergence of stable patterns of causal interactions among cortical areas during NREM.

A particularly intriguing possibility is that the occurrence of bistable dynamics may be relevant also for brain-injured patients in which consciousness is lost in spite of preserved neuronal activity [42]. Brain lesions may in fact indirectly induce sleep-like bistability in intact portions of the cortical mantle in several ways. As an example, a direct damage occurring at the level of brainstem activating systems (reducing/abolishing the cholinergic, noradrenergic, histaminergic, and glutamatergic ascending drive) may result in enhanced K+ conductances at the cortical level [43]. On the other hand, localized cortical and subcortical alterations (e.g., undetected epileptic foci) may affect the function of brainstem activating systems and, in turn, produce diffuse cortical bistability [44]. Another possibility is represented by the occurrence of cortical deafferentation following white matter damage induced by the lesion. In these respects, animal experiments in which white matter was severed with a cortical undercut resulted in slow waves and in a continuous alternation between up and down states in the partially deafferented gyrus, even when the animal was behaviorally awake and the rest of the brain was displaying typical wake EEG patterns [45]. Alternatively, a form of bistability may result from cortical and subcortical lesions that alter the cortical balance between excitation and inhibition in favor of inhibition in intact portions of the thalamocortical system; this excess in inhibition may be either generated locally or projected by healthy areas that become hyperactive [46]. This possibility is conceivable given the pharmacologically induced loss of consciousness observed when employing GABAergic agents such as midazolam [32]. Finally, a reduction of the excitatory drive from dopaminergic striatal and frontostriatal circuits on cortical neurons due to subcortical lesions involving the central thalamic nuclei [47] may lead to sleep-like bistability in diffuse cortical circuits and may offer a key to explain the paradoxical effects of the sedative zolpidem, a non-benzodiazepine hypnotic that potentiates GABAA receptors, on behavioral improvement of alertness and interactive behavior in severely brain-injured patients [48, 47, 49]. The role of sleep-like bistability in VS/UWS

Fig. 8.2 During NREM, perturbations trigger a slow-wave-like response that is associated with high-frequency suppression followed by decay of PLF. Panel (**a**) shows the average TMS-evoked potential (TEP) recorded in a healthy individual from one channel located under the stimulator during wakefulness (W—*red*) and NREM sleep (NREM—*blue*). In panel (**b**) the following measures are reported for one representative intracerebral (stereo-EEG) bipolar contact. *CCEPs*: the average response following intracerebral electrical stimulation during wakefulness (W—*red*) and NREM (NREM—*blue*). *ERSPs:* time-frequency power spectra of CCEPs recorded in W and NREM. *Blue color* indicates a significant reduction compared to the baseline, while *red indicates* significant increase. The *dashed horizontal line* indicates 20 Hz. PWR >20 Hz: time series of high-frequency power (>20 Hz). *PLF >8 Hz:* PLF calculated on a single trial level both for W (*red*) and NREM (*blue*) after high-pass filtering (>8 Hz). *Dashed vertical lines* and triangles represent stimulus onset (Modified from Pigorini et al. [35])

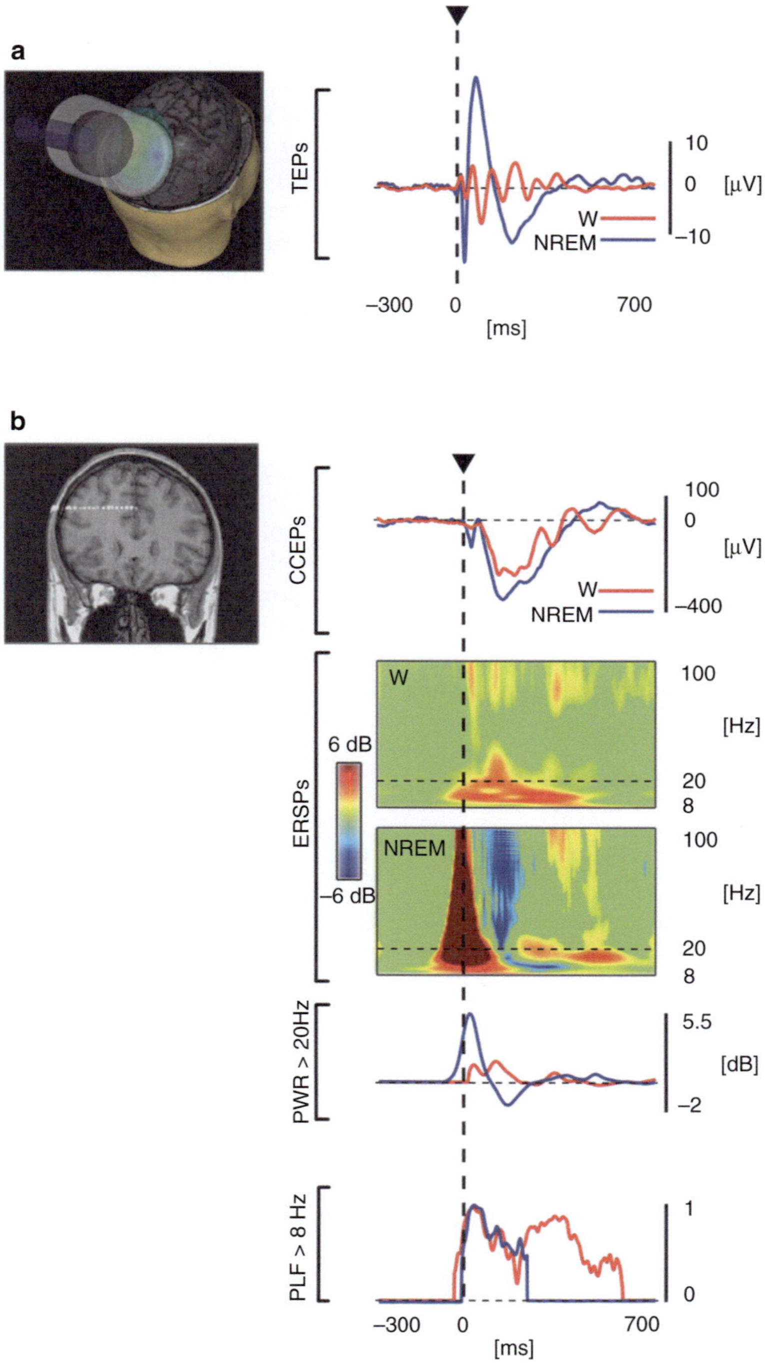
a
TEPs
10
0 [μV]
–10
W
NREM
–300 0 700
[ms]
b
CCEPs
100
0
[μV]
–400
W
NREM
ERSPs
6 dB
–6 dB
W
NREM
100
[Hz]
20
8
PWR > 20Hz
5.5
[dB]
–2
PLF > 8 Hz
1
0
–300 0 700
[ms]

patients is in line with the presence of low-frequency activity often reported as a prominent feature of the EEG spectrum in these patients [50].

In all cases, an intriguing possibility is that intrinsic bistability affecting portions of the thalamocortical system may per se prevent the emergence of stable patterns of TMS-related deterministic interactions among cortical areas. Evaluating the presence of bistability in the cerebral cortex of brain-injured patients would thus be critically important, especially if one considers that 30 % of clinically VS/UWS show preserved metabolic levels in the frontoparietal cortex [51, 42]. In these cases, just like during NREM sleep, bistability may prevent the emergence of complex dynamics in spite of preserved structural connections and levels of activity. Hence, the intriguing perspective that in VS/UWS patients it may still be possible to reduce bistability and increase network complexity by acting pharmacologically on intrinsic neuronal properties. Experimentally, one should first demonstrate that the slow-wave-like graphoelement triggered by TMS in VS/UWS patients truly reflects a neuronal down state (as indexed by a concomitant suppression of EEG power above 20 Hz) and that the following spectral activity is not phase-locked with the TMS pulse. Then, one should ascertain the causal role of bistability by showing that a reduction of the TMS-triggered down state heralds the recovery of consciousness in acute patients evolving from coma to different degrees of clinical recovery. In turn, this should be paralleled by a recovery of long-range, deterministic complex activations at the EEG level [33]. Both these observations would in fact strongly suggest that bistability may be the common final pathway for a collapse of brain complexity causing LOC.

References

1. Cruse D, Chennu S, Chatelle C, Bekinschtein TA, Fernandez-Espejo D, Pickard JD et al (2011) Bedside detection of awareness in the vegetative state: a cohort study. Lancet 378(9809):2088–2094. doi:10.1016/S0140-6736(11)61224-5
2. Naci L, Cusack R, Anello M, Owen AM (2014) A common neural code for similar conscious experiences in different individuals. Proc Natl Acad Sci U S A 111(39):14277–14282. doi:10.1073/pnas.1407007111
3. Owen AM, Coleman MR, Boly M, Davis MH, Laureys S, Pickard JD (2006) Detecting awareness in the vegetative state. Science 313(5792):1402
4. Bardin JC, Fins JJ, Katz DI, Hersh J, Heier LA, Tabelow K et al (2011) Dissociations between behavioural and functional magnetic resonance imaging-based evaluations of cognitive function after brain injury. Brain 134(Pt 3):769–782. doi:10.1093/brain/awr005
5. Koulack D (1969) Effects of somatosensory stimulation on dream content. Arch Gen Psychiatry 20(6):718–725
6. Oizumi M, Albantakis L, Tononi G (2014) From the phenomenology to the mechanisms of consciousness: integrated information theory 3.0. PLoS Comput Biol 10(5):e1003588. doi:10.1371/journal.pcbi.1003588
7. Dehaene S, Changeux JP (2011) Experimental and theoretical approaches to conscious processing. Neuron 70(2):200–227. doi:10.1016/j.neuron.2011.03.018
8. Friston K (2002) Beyond phrenology: what can neuroimaging tell us about distributed circuitry? Annu Rev Neurosci 25:221–250. doi:10.1146/annurev.neuro.25.112701.142846

9. Laureys S (2005) The neural correlate of (un)awareness: lessons from the vegetative state. Trends Cogn Sci 9(12):556–559. doi:10.1016/j.tics.2005.10.010
10. Tononi G, Koch C (2008) The neural correlates of consciousness: an update. Ann N Y Acad Sci 1124:239–261. doi:10.1196/annals.1440.004
11. Seth AK, Izhikevich E, Reeke GN, Edelman GM (2006) Theories and measures of consciousness: an extended framework. Proc Natl Acad Sci U S A 103(28):10799–10804. doi:10.1073/pnas.0604347103
12. Sporns O (2011) The human connectome: a complex network. Ann N Y Acad Sci 1224: 109–125. doi:10.1111/j.1749-6632.2010.05888.x
13. Tononi G (2004) An information integration theory of consciousness. BMC Neurosci 5:42. doi:10.1186/1471-2202-5-42
14. Tononi G, Edelman GM (1998) Consciousness and complexity. Science 282(5395): 1846–1851
15. Boly M (2011) Measuring the fading consciousness in the human brain. Curr Opin Neurol 24(4):394–400. doi:10.1097/WCO.0b013e328347da94
16. Seth AK, Dienes Z, Cleeremans A, Overgaard M, Pessoa L (2008) Measuring consciousness: relating behavioural and neurophysiological approaches. Trends Cogn Sci 12(8):314–321
17. G. Tononi, O. Sporns, G. M. Edelman, A measure for brain complexity: Relating functional segregation and integration in the nervous system. Proc. Natl. Acad. Sci. U.S.A. 91, 5033–5037 (1994)
18. G. Tononi, C. Koch, The neural correlates of consciousness: An update. Ann. N. Y. Acad. Sci. 1124, 239–261 (2008)
19. A. K. Seth, A. B. Barrett, L. Barnett, Causal density and integrated information as measures of conscious level. Philos. Trans. A Math. Phys. Eng. Sci. 369, 3748–3767 (2011)
20. Engel AK, Singer W (2001) Temporal binding and the neural correlates of sensory awareness. Trends Cogn Sci 5(1):16–25
21. Kotchoubey B (2005) Apallic syndrome is not apallic: is vegetative state vegetative? Neuropsychol Rehabil 15(3–4):333–356
22. Sitt JD, King JR, El Karoui I, Rohaut B, Faugeras F, Gramfort A et al (2014) Large scale screening of neural signatures of consciousness in patients in a vegetative or minimally conscious state. Brain 137(Pt 8):2258–2270. doi:10.1093/brain/awu141
23. Johnson, R. W. (1987). Relative-entropy minimization with uncertain constraints: theory and application to spectrum analysis (pp. 57-73). Springer Netherlands.
24. Pincus SM, Gladstone IM, Ehrenkranz RA (1991) A regularity statistic for medical data analysis. J Clin Monit 7(4):335–345
25. Massimini M, Boly M, Casali A, Rosanova M, Tononi G (2009) A perturbational approach for evaluating the brain's capacity for consciousness. Prog Brain Res 177:201–214. doi:10.1016/S0079-6123(09)17714-2
26. Ilmoniemi RJ, Virtanen J, Ruohonen J, Karhu J, Aronen HJ, Naatanen R et al (1997) Neuronal responses to magnetic stimulation reveal cortical reactivity and connectivity. Neuroreport 8(16):3537–3540
27. Casali AG, Gosseries O, Rosanova M, Boly M, Sarasso S, Casali KR et al (2013) A theoretically based index of consciousness independent of sensory processing and behavior. Sci Transl Med 5(198):198ra10. doi:10.1126/scitranslmed.300
28. Giacino JT, Kalmar K, Whyte J (2004) The JFK Coma Recovery Scale-Revised: measurement characteristics and diagnostic utility. Arch Phys Med Rehabil 85(12):2020–2029
29. Massimini M, Ferrarelli F, Huber R, Esser SK, Singh H, Tononi G (2005) Breakdown of cortical effective connectivity during sleep. Science 309(5744):2228–2232. doi:10.1126/science.1117256
30. Rosanova M, Casali A, Bellina V, Resta F, Mariotti M, Massimini M (2009) Natural frequencies of human corticothalamic circuits. J Neurosci 29(24):7679–7685. doi:10.1523/JNEUROSCI.0445-09.2009
31. Massimini M, Ferrarelli F, Esser SK, Riedner BA, Huber R, Murphy M et al (2007) Triggering sleep slow waves by transcranial magnetic stimulation. Proc Natl Acad Sci U S A 104(20):8496–8501. doi:10.1073/pnas.0702495104

32. Ferrarelli F, Massimini M, Sarasso S, Casali A, Riedner BA, Angelini G et al (2010) Breakdown in cortical effective connectivity during midazolam-induced loss of consciousness. Proc Natl Acad Sci U S A 107(6):2681–2686. doi:10.1073/pnas.0913008107
33. Rosanova M, Gosseries O, Casarotto S, Boly M, Casali AG, Bruno MA et al (2012) Recovery of cortical effective connectivity and recovery of consciousness in vegetative patients. Brain 135(Pt 4):1308–1320. doi:10.1093/brain/awr340
34. Massimini M, Ferrarelli F, Murphy M, Huber R, Riedner B, Casarotto S et al (2010) Cortical reactivity and effective connectivity during REM sleep in humans. Cogn Neurosci 1(3):176–183. doi:10.1080/17588921003731578
35. Steriade M, Nunez A, Amzica F (1993) A novel slow (<1 Hz) oscillation of neocortical neurons in vivo: depolarizing and hyperpolarizing components. J Neurosci 13(8):3252–3265
36. Compte A, Sanchez-Vives MV, McCormick DA, Wang XJ (2003) Cellular and network mechanisms of slow oscillatory activity (<1 Hz) and wave propagations in a cortical network model. J Neurophysiol 89(5):2707–2725. doi:10.1152/jn.00845.2002
37. Sanchez-Vives MV, McCormick DA (2000) Cellular and network mechanisms of rhythmic recurrent activity in neocortex. Nat Neurosci 3(10):1027–1034. doi:10.1038/79848
38. Pigorini A, Sarasso S, Proserpio P, Szymanski C, Arnulfo G, Casarotto S et al (2015) Bistability breaks-off deterministic responses to intracortical stimulation during non-REM sleep. Neuroimage 112:105–113. doi:10.1016/j.neuroimage.2015.02.056
39. Cash SS, Halgren E, Dehghani N, Rossetti AO, Thesen T, Wang C et al (2009) The human K-complex represents an isolated cortical down-state. Science 324(5930):1084–1087. doi:10.1126/science.1169626
40. Valderrama M, Crepon B, Botella-Soler V, Martinerie J, Hasboun D, Alvarado-Rojas C et al (2012) Human gamma oscillations during slow wave sleep. PLoS One 7(4), e33477. doi:10.1371/journal.pone.0033477
41. Sinkkonen J, Tiitinen H, Naatanen R (1995) Gabor filters: an informative way for analysing event-related brain activity. J Neurosci Methods 56(1):99–104
42. Stender J, Gosseries O, Bruno MA, Charland-Verville V, Vanhaudenhuyse A, Demertzi A et al (2014) Diagnostic precision of PET imaging and functional MRI in disorders of consciousness: a clinical validation study. Lancet 384(9942):514–522. doi:10.1016/S0140-6736(14)60042-8
43. McCormick DA, Wang Z, Huguenard J (1993) Neurotransmitter control of neocortical neuronal activity and excitability. Cereb Cortex 3(5):387–398
44. Englot DJ, Yang L, Hamid H, Danielson N, Bai X, Marfeo A et al (2010) Impaired consciousness in temporal lobe seizures: role of cortical slow activity. Brain 133(Pt 12):3764–3777. doi:10.1093/brain/awq316
45. Timofeev I, Grenier F, Bazhenov M, Sejnowski TJ, Steriade M (2000) Origin of slow cortical oscillations in deafferented cortical slabs. Cereb Cortex 10(12):1185–1199
46. Murase N, Duque J, Mazzocchio R, Cohen LG (2004) Influence of interhemispheric interactions on motor function in chronic stroke. Ann Neurol 55(3):400–409. doi:10.1002/ana.10848
47. Schiff ND (2010) Recovery of consciousness after brain injury: a mesocircuit hypothesis. Trends Neurosci 33(1):1–9. doi:10.1016/j.tins.2009.11.002
48. Brefel-Courbon C, Payoux P, Ory F, Sommet A, Slaoui T, Raboyeau G et al (2007) Clinical and imaging evidence of zolpidem effect in hypoxic encephalopathy. Ann Neurol 62(1):102–105. doi:10.1002/ana.21110
49. Whyte J, Myers R (2009) Incidence of clinically significant responses to zolpidem among patients with disorders of consciousness: a preliminary placebo controlled trial. Am J Phys Med Rehabil 88(5):410–418. doi:10.1097/PHM.0b013e3181a0e3a0
50. Schiff ND, Nauvel T, Victor JD (2014) Large-scale brain dynamics in disorders of consciousness. Curr Opin Neurobiol 25:7–14. doi:10.1016/j.conb.2013.10.007
51. Gosseries O, Di H, Laureys S, Boly M (2014) Measuring consciousness in severely damaged brains. Annu Rev Neurosci 37:457–478. doi:10.1146/annurev-neuro-062012-170339

Chapter 9
Brain Electrophysiology in Disorders of Consciousness: Diagnostic and Prognostic Utility

Damian Cruse, G. Bryan Young, Francesco Piccione, Marianna Cavinato, and Aldo Ragazzoni

Abstract Electroencephalography (EEG) is a powerful and inexpensive bedside tool for the assessment of residual brain function in prolonged disorders of consciousness. Here we review a range of methods for EEG interpretation, including reactivity, perturbation by transcranial magnetic stimulation, evoked potentials, and oscillatory changes. We show that, in combination, these methods can form a reliable picture of each patient's structural impairments and residual sensory and cognitive functioning, thereby leading to more accurate diagnoses and prognoses and stratification of patients for experimental neuro-stimulation therapies.

Keywords Disorders of consciousness • EEG • ERP • TMS • Diagnosis • Prognosis

9.1 Introduction

For many decades, electroencephalography (EEG) has provided clinicians with a bedside tool to assess brain function following severe brain injury. The advent of modern computing has greatly increased the range of inferences that can be drawn from EEG data, thereby providing both diagnostic and prognostic utility. Here we

D. Cruse (✉)
The Brain and Mind Institute, University of Western Ontario, London, ON, Canada
e-mail: dcruse@uwo.ca

G.B. Young
Department of Clinical Neurological Sciences, University of Western Ontario, London, ON, Canada

F. Piccione • M. Cavinato
I.R.C.C.S, San Camillo Hospital, Venezia, Italy

A. Ragazzoni
Unit of Neurology and Clinical Neurophysiology, Fondazione PAS, Scandicci, Firenze, Italy

M.M. Monti, W.G. Sannita (eds.), *Brain Function and Responsiveness in Disorders of Consciousness*, DOI 10.1007/978-3-319-21425-2_9

review the range of approaches to EEG assessments of patients with prolonged disorders of consciousness (pDOC) – i.e., vegetative state (VS)/unresponsive wakefulness state (UWS) and minimally conscious state (MCS).

9.2 EEG Reactivity

A change in EEG amplitude and/or frequency within a few seconds of a stimulus is referred to as "reactivity" [1]. Stimuli are usually auditory (hand claps, shouting the patient's name), nociceptive (nail bed pressure, nasal tickle), or passive eyelid opening. EEG reactivity to a stimulus reflects an interaction between brainstem reticular arousal systems and cerebral structures that support corticothalamic interactions [2]. Reactivity during acute DOC (i.e., coma) is predictive of survival/recovery [1, 3–7] although not perfectly so [8]. However, there have been relatively few studies of EEG reactivity in patients with pDOC.

Noxious stimuli have failed to provoke visually interpreted EEG reactivity in a group of UWS patients [9], while patients with pDOCs who exhibited EEG reactivity also exhibited significantly higher behavioral abilities [10]. Furthermore, those patients with EEG reactivity also showed significantly greater behavioral improvement at 3-month follow-up. Ragazzoni and colleagues [11] used transcranial magnetic stimulation (TMS) over the motor cortex to induce EEG reactivity and found variable reactivity over the ipsilateral cortex and absent reactivity over the contralateral cortex in UWS patients, while MCS patients exhibited relatively preserved but abnormal responses (see Fig. 9.1).

The primary benefit of assessing EEG reactivity is that it is easy to perform and can be interpreted by visual inspection from a trained EEG reviewer. However, the subjective and time-consuming nature of the analysis and inconsistency among reviewers (depending on experience and expertise) has prompted some early attempts at quantitative analysis [12–15]. Although quantitative measures provide more objectivity to the assessment of reactivity, they are vulnerable to artifactual EEG changes that might be mistaken as EEG reactivity and should be corroborated by experienced reviewers.

9.3 Short-Latency Evoked Potentials

Short-latency evoked potentials (EPs) are stimulus-locked EEG responses that occur within a brief interval from a stimulus (~10–150 ms) and are shaped by the physical characteristics of that stimulus. Together they measure the integrity of the central sensory pathways from the brainstem to the primary sensory cortices and thereby provide a crucial measure of basic cortical integrity. Somatosensory EPs (SEPs), for example, are unanimously acknowledged as reliable predictors of positive/negative outcome in acute DOC [16–18] and have been included in major practice guidelines [4, 19]. However, short-latency EPs have been the object of a limited number of studies in pDOC.

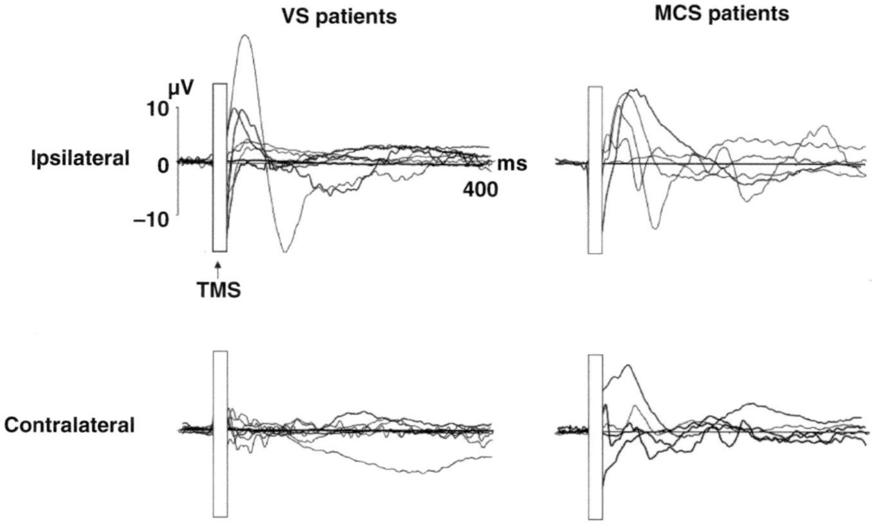

Fig. 9.1 Individual cortical potentials evoked by transcranial magnetic stimulation (*TEPs*) of the primary motor cortex and recorded from both hemispheres for the two groups of patients: VS patients (8) are shown in the *left column*, and MCS patients (5) are shown in the *right column*. TEPs recorded over the stimulated hemisphere are displayed in the upper row, and TEPs recorded over the contralateral hemisphere are in the lower row (Reproduced under the terms of a Creative Commons Attribution License from Ragazzoni et al. 2013 [11])

Brainstem auditory EPs are usually normal or delayed, with no differences between UWS and MCS [9, 20–22] and no prognostic implications [23, 24]. Bilateral abolition of cortical SEP components (N20 and the following) is a frequent finding in UWS following anoxia [25], at variance with posttraumatic patients [23]. Ragazzoni et al. [11] found no correlation between SEP abnormalities and clinical diagnosis. However, the presence of median nerve cortical SEPs reliably predicts long-term recovery of responsiveness in anoxic patients [25]. Visual EP latencies can also predict long-term outcome in the post-acute UWS [26, 27].

Short-latency EPs therefore provide crucial information on the structural integrity of ascending pathways and are easily employed as an extension of the neurological examination. However, their power in making the differential diagnosis between UWS and MCS is weak.

9.4 Event-Related Potentials

Event-related potentials (ERPs) represent a class of EEG responses with latencies longer than 100 ms that reflect the mass activity of neuron assemblies underpinning a series of cognitive processes. The sequence and latencies of ERP components are related to successive stages of the information-processing stream, spanning from perception to higher-order cerebral processes, such as attention, memory updating, and semantic comprehension. Importantly, the earlier components elicited in the 100–250 ms time interval (such as P1, N1, P2, N2, mismatch negativity—MMN) have been associated with the automatic sensory and perceptual processing of stimuli, operating independently from attention on an unconscious level.

Access to conscious awareness is signaled by the appearance over the central-parietal scalp of a later positive component, named P3b (or P300 for its modal latency) peaking between 300 and 500 ms, depending on the sensory modality of the eliciting stimulus [28–30]. Although not all experimental evidences support this interpretation (for a detailed discussion [31]), P3b appears at present as one of the most reliable electrophysiological marker of conscious access.

An abundance of ERP studies in pDOC have been produced in the last 20 years. Some reports concerned single or few patients or were cohort studies and are therefore of limited clinical relevance and will not be considered here. Table 9.1 summarizes the results of thirteen studies with relatively large samples of patients analyzed at the individual level and aiming at differentiating patients with UWS from those with MCS. We categorize the studies into two levels, reflecting the different complexities of ERP components analyzed. Level 1 indicates analysis of earlier components (i.e., MMN, P3a, N400) associated with preconscious cognitive functions, such as sensory memory, orienting response, or semantic processing. Level 2 refers to analysis of later components (P3b, LPC, P600) that in healthy controls reflect conscious processing.

Inspection of Table 9.1 indicates that, remarkably, an average of 1 in 3 UWS patients presented ERP responses at level 1 and/or 2, indicating that a substantial proportion of the patients could harbor some level of cognitive processing and even of conscious awareness. Unfortunately, the clinical evidence of a recovery of consciousness was available only for a few studies. Another notable fact is that no more than 43 % of MCS patients presented with ERP signs of consciousness, despite behavioral evidence of conscious awareness. Possible explanations for these false-negative results are fluctuations in arousal, lack of motivation, fatigue, difficulty in understanding the task, sensory defects, technical factors such as artifacts, and latency variability of responses. These arguments a fortiori are valid for UWS patients and suggest that ERPs underestimate the cognitive capacities of DOC patients and have poor power for differential diagnosis.

The prognostic value of ERPs has been tested mostly in terms of recovery of consciousness and neurological functions. Table 9.2 presents results from the main seven studies addressing prognosis in DOC, with follow-ups extending to 14 years. The strongest ERP predictor for recovery of consciousness is the presence of MMN, which also provided prognostic information on the functional outcome. P3b emerged as a less reliable predictor of outcome, whereas the presence of N400 was associated with a favorable neurological recovery over a long-term follow-up period. There is therefore growing evidence for the prognostic power of ERPs in pDOC.

Table 9.1 ERP diagnostic studies in UWS and MCS

Authors	ERP examined	UWS N°	Response level 1[a]	Response level 2[b]	MCS N°	Response level 1[a]	Response level 2[b]	Task
Kotchoubey et al. (2005) [68]	MMN, P3b, N400, P600	50	26 (52 %)	12 (24 %)	38	13 (34 %)	14 (36 %)	Auditory and semantic oddball
Perrin et al. (2006) [69]	N1, P3b	5		3 (60 %)	6		6 (100 %)	SON oddball
Schnakers et al. (2008) [70]	N1, P3b	8		0 (0 %)	14		9 (64 %)	SON oddball
Bekinschtein et al. (2009) [71]	MMN, P3b	4	3 (75 %)	0 (0 %)	4	4 (100 %)	3 (75 %)	Auditory local-global paradigm
Fischer et al. (2010) [22]	MMN, P3a, P3b	16	3 (19 %)	1 (6 %)	11	4 (36 %)	1 (9 %)	Auditory and SON oddball
Cavinato et al. (2011) [72]	N1, P3b	11		5 (45 %)	6		6 (100 %)	Auditory and SON oddball
Boly et al. (2011) [73]	MMN	8	8 (100 %)		13	13 (100 %)		Auditory oddball
Holler et al. (2011) [74]	MMN	16	2 (12 %)		6	0 (0 %)		Auditory oddball
Faugeras et al. (2012) [75]	MMN, P3b	24	6 (25 %)	2 (8 %)	28	9 (32 %)	4 (14 %)	Auditory local-global paradigm
Chennu et al. (2013) [76]	P3a, P3b	9	1 (11 %)	1 (11 %)	12	3 (25 %)	0 (0 %)	Auditory verbal oddball
Ragazzoni et al. (2013) [11]	N1, P3b	8		0 (0 %)	5		0 (0 %)	Auditory oddball
Steppacher et al. (2013) [77]	N400, P3b	50	19 (38 %)	32 (64 %)	39	18 (46 %)	29 (74 %)	Auditory oddball, sentences
Rohaut et al. (2015) [78]	N400, LPC	15	1 (7 %)	1 (7 %)	14	5 (36 %)	5 (36 %)	Auditory semantic priming paradigm
Total number		224	69 (34 %)	57 (28 %)	196	69 (42 %)	77 (43 %)	

Abbreviations: *N°* number of patients, *UWS* vegetative state/unresponsive wakefulness syndrome, *MCS* minimally conscious state, *MMN* mismatch negativity, *LPC* late positive component, *SON* subject's own name, *%* percentage of patients in which ERP components were detected over the total number of patients examined for the exact same ERPs

[a]Response level 1 indicates detection of ERP components MMN, P3a, and N400

[b]Response level 2 indicates detection of ERP components P3b and LPC (P600)

Table 9.2 ERP prognostic studies in UWS and MCS

Authors	ERP examined	UWS N°	MCS N°	Follow-up	
Kotchoubey et al. (2005) [68]	MMN	23	20	6 mos	MMN absent: 22 % of patients improved MMN present: 59 % of patients improved
Wijnen et al. (2007) [79]	MMN	10		1.5–5.2 mos	MMN amplitude and latency predict recovery from VS
Qin et al. (2008) [80]	MMN	6	2	3 mos	SON MMN presence predicts recovery of consciousness
Cavinato et al. (2009) [23]	P3b	34		1 yr	P300 presence predicts recovery of consciousness
Luauté et al. (2010) [24]	MMN	12	39	5 yrs	MMN not associated with favorable outcome
Steppacher et al. (2013) [77]	N400, P3b	50	39	2–14 yrs	N400 presence predicts favorable outcome P3b presence does not predict outcome
Wijnen et al. (2014) [81]	N200, P3b, N350	10		1.5–5.2 mos	Presence and amplitude of N350 predict outcome
Total number		145	100		

Abbreviations: *UWS* vegetative state/unresponsive wakefulness syndrome, *MCS* minimally conscious state, *MMN* mismatch negativity, *MOS* months, *yr/yrs* year/years, *SON* subject's own name

9.5 Cognitive EEG Oscillations

Following the success of fMRI mental imagery paradigms to identify patients misdiagnosed as in the UWS, some success has been achieved through similar approaches in EEG. Using a variant of the tennis imagery task from fMRI [32], Goldfine et al. [33] observed significant increases in posterior alpha rhythms when one MCS patient was instructed to imagine swimming, relative to when they were instructed to rest. Furthermore, at 4-month follow-up, this patient had emerged from the MCS.

However, as noted by Goldfine et al. [33] and others [34, 35], even within healthy individuals, there is a high level of variability in the type of EEG modulation associated with mental imagery. Methods from "machine learning" have therefore been employed in order to identify the subject-specific EEG response that is consistent during periods of mental imagery, which do not rely on strict a priori predictions.

In the first example of this approach, Cruse et al. [36] asked a group of 16 UWS patients to alternately imagine moving their right hands and toes. By analyzing the spatiotemporal power values within EEG frequency bands known to be associated with motor imagery, support vector machines were able to significantly differentiate between the two types of motor imagery in 75 % of healthy individuals. Crucially, significantly greater than chance differentiation was also observed in 3 out of the 16

VS patients, indicating that these patients were following the commands and had therefore been misdiagnosed. Using the same approach, 3 out of 8 MCS patients who were incapable of following commands with their external behavior showed signs of covert command following in their EEG [37].

Despite the strengths of machine-learning methods for identifying subject-specific responses, they are also susceptible to EEG changes that are not task-relevant. For example, in a documented discussion of statistical methods [38–40], Goldfine et al. [39] highlighted the potential problems of blocked mental imagery tasks, such as those employed by Cruse et al. [36]. A blocked design overcomes criticisms of the potential automatic activation caused by task instructions [41] but also introduces potential violations of the assumptions of certain classes of statistical methods that may lead to unacceptable rates of false positives. While increasing the amount of EEG data features that are available for machine learning can in fact decrease the rate of false positives [42], the accuracy of a diagnosis made on the basis of neuroimaging-detected command following is of great importance. As has been discussed extensively elsewhere [43, 44], in all statistical analyses, a case-by-case trade-off must be drawn between an acceptable rate of false positives (i.e., evidence of command following when the patient is in fact unaware) and an acceptable rate of misses (i.e., no evidence of command following when the patient is in fact aware).

Nevertheless, in subsequent years, alternative methods that were demonstrably less susceptible to false positives have found EEG command-following abilities in a number of UWS patients (see Fig. 9.2; [34, 45]). Furthermore, Horki et al. [46] found that imagery of a user-selected sport was sufficient to produce appropriate modulations of EEG rhythms in a group of MCS patients. Oscillatory measures of command following therefore have shown promise for diagnosing awareness in pDOC. However, their prognostic value is currently unknown.

9.6 Sleep EEG

The return of "wakefulness" is a diagnostic characteristic of entry into a pDOC. However, the results of the limited EEG investigations of sleep-wake cycling in pDOC patients are inconsistent with the assumption of preserved wakefulness in UWS patients [47]. This fact, however, can be employed as a diagnostic marker. Isono et al. [48] reported no quantifiable EEG changes between overnight and daytime periods in four out of twelve UWS patients. Landsness et al. [49] observed sleep-wake-like changes in the EEG of six MCS patients across 1 day, while the EEG of five UWS patients remained unchanged between periods of eye opening and eye closure. Furthermore, sleep spindles and/or REM sleep was observed in all MCS patients but was not observed in any of the VS patients.

Further emphasizing the diagnostic value of assessments of sleep architecture, De Biase et al. [50] found that patients showing evidence of both REM sleep and sleep spindles scored significantly higher on behavioral scales of awareness than

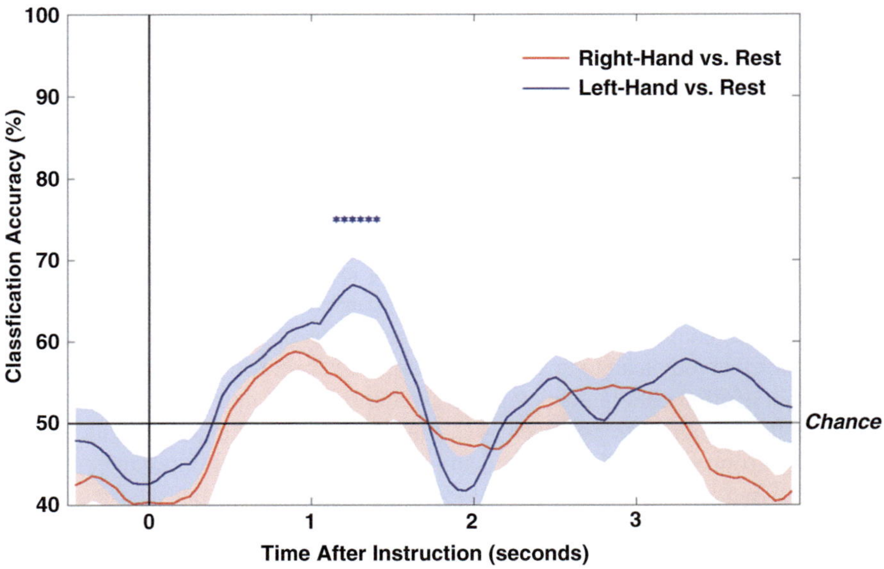

Fig. 9.2 Evidence for covert command following in a patient diagnosed as VS/UWS for more than 12-years. Data show time courses of machine-learning discrimination (classification accuracy) between EEG activity at rest and EEG activity following instructions to move the right and left hands. Stars denote time points with significantly above-chance discrimination for left hand vs. rest ($p<025$) (Reproduced under the terms of a Creative Commons Attribution License from Cruse et al. 2012 [34])

those patients for whom both of these sleep markers were absent. Sleep spindles are a characteristic of non-REM sleep and reflect interactions between the thalamus and cortex. The absence of sleep spindles in behaviorally less-responsive patients is consistent with structural and functional data that find specifically impaired thalamocortical connectivity in these patients [51, 52].

It is evident that sleep EEG provides both diagnostic information and an indirect measure of neural connectivity in pDOC. Furthermore, sleep-wake assessments can identify times of day in which the patient may be more alert and therefore more likely to respond to behavioral prompts or therapies.

9.7 Putting Consciousness Theory into Practice

While many EEG approaches to identifying residual consciousness employ methods akin to conventional behavioral assessments – i.e., evidence for command following is considered evidence for consciousness – in recent years there has been a growth in theoretically based markers of awareness. With the increased availability of neuroimaging, a number of theories of consciousness have gained support and made specific predictions regarding pDOC. One set of theories considers the ability of the brain to both integrate and differentiate information to be a key mechanism for supporting human consciousness [29, 53].

One experimental approach has been to assess EEG responses to transcranial magnetic stimulation (TMS). In awake UWS patients, TMS triggers a simple, local slow response that indicates a breakdown in effective connectivity, similar to that observed in unconscious sleeping or anesthetized patients [54–56]. In contrast, in MCS patients, TMS triggers complex activations that sequentially involve distant cortical areas ipsilateral and contralateral to the site of stimulation. Casali et al. [57] quantified the complexity of EEG responses evoked by TMS with the perturbational complexity index (PCI), a measure of the amount of information integration and differentiation within the brain's response to perturbation. Consistent with the theory linking integration and differentiation to consciousness, PCI values were consistently lower during sleep and sedation than during wakeful periods. Interestingly, PCI values also tracked the behavioral signs of awareness shown by DOC patients, with values for UWS within the bounds of sleep and anesthesia, and those for MCS between healthy wakefulness and sleep/sedation, indicating the potential diagnostic value of this index.

Repetitive TMS can also induce significant perturbations on subcortical regions, including increased dopamine release and long-lasting oscillatory effects, indicating a potential therapeutic effect [58, 59]. rTMS modulates the excitability of targeted cerebral areas alongside changes in remote cortical and subcortical regions, including the thalamus and striatum, and has been shown to modify motor, sensory, and high-level cognitive functions [60, 61]. Indeed, a 10 Hz rTMS protocol applied to a MCS patient promoted clinically significant neurobehavioral recovery [62].

Noninvasive rTMS may therefore provide a new tool for brain investigation and could identify those patients most likely to benefit from therapy. Indeed, Strafella et al. [63] found that rTMS of the motor cortex induced the release of striatal dopamine. There is mounting evidence that consciousness depends on the capacity of remote brain regions to interact through cortico-cortical and cortico-thalamo-cortical connections [51]. The central thalamus and cortex are closely linked through both direct connections and indirect connections through the frontal cortical striato-pallidal-thalamocortical loops. Modulations of striatal activity by cortical rTMS may therefore help strengthen these pathways and restore awareness.

As a measure of neural response to auditory stimuli, King et al. [64] introduced the weighted symbolic mutual information (wSMI) measure to evaluate the extent to which two EEG signals fluctuate together in a nonrandom fashion. Again, consistent with theories of consciousness, this measure of information sharing across the brain reliably differentiated UWS patients from MCS patients, who in turn were differentiated from fully conscious individuals.

Chennu et al. [65] calculated connectivity indices within canonical frequency bands of resting state EEG recordings and performed graph theoretical analyses in order to quantify the structure of these resting networks. Specifically it was found that long-distance connectivity within the alpha band (8–12 Hz) was greatly reduced in both UWS and MCS patients. Interestingly, the networks exhibited by UWS patients who could perform mental imagery to command in the fMRI were similar to those seen in healthy awake individuals. Furthermore, an increase in the alpha band of a single MCS patient was correlated with functional improvement [66].

9.8 Conclusions

No single approach to the diagnosis and prognosis of pDOC is perfectly sensitive. Indeed, while corroborations have been observed across modalities, a number of patients have been reported for whom only one method has brought their behavioral diagnosis into question [45, 67]. Due to the fluctuations in behavioral abilities across time, and the heterogeneity of brain injuries that characterize pDOC, a range of markers of awareness that are produced by spatially distinct brain regions and detected with multiple modalities are required to accurately determine a patient's true level of awareness and potential for recovery.

References

1. Gutling E, Gonser A, Imhof HG, Landis T (1995) EEG reactivity in the prognosis of severe head injury. Neurology 45:915–918. doi:10.1212/WNL.45.5.915
2. Kobylarz E, Schiff N (2005) Neurophysiological correlates of persistent vegetative and minimally conscious states. Neuropsychol Rehabil 15:323–332. doi:10.1080/09602010443000605
3. Thenayan Al E, Savard M, Sharpe M et al (2008) Predictors of poor neurologic outcome after induced mild hypothermia following cardiac arrest. Neurology 71:1535–1537. doi:10.1212/01.wnl.0000334205.81148.31
4. Guerit JM, Amantini A, Amodio P et al (2009) Consensus on the use of neurophysiological tests in the intensive care unit (ICU): Electroencephalogram (EEG), evoked potentials (EP), and electroneuromyography (ENMG). Neurophysiol Clin 39:71–83. doi:10.1016/j.neucli.2009.03.002
5. Kaplan PW, Genoud D, Ho TW, Jallon P (2000) Clinical correlates and prognosis in early spindle coma. Clin Neurophysiol 111:584–590. doi:10.1016/S1388-2457(99)00303-X
6. RamachandranNair R, Sharma R, Weiss SK, Cortez MA (2005) Reactive EEG Patterns in Pediatric Coma. Pediatr Neurol 33:345–349. doi:10.1016/j.pediatrneurol.2005.05.007
7. Rossetti AO, Urbano LA, Delodder F, Kaplan PW (2010) Prognostic value of continuous EEG monitoring during therapeutic hypothermia after cardiac arrest. Crit Care 14(5):R173
8. Logi F, Pasqualetti P, Tomaiuolo F (2011) Predict recovery of consciousness in post-acute severe brain injury: the role of EEG reactivity. Brain Inj 25:972–979. doi:10.3109/02699052.2011.589795
9. Hansotia PL (1985) Persistent vegetative state. Arch Neurol 42:1048–1052. doi:10.1001/archneur.1985.04060100030015
10. Bagnato S, Boccagni C, Sant'Angelo A et al (2015) EEG predictors of outcome in patients with disorders of consciousness. Clin Neurophysiol 126(5):959–966. doi:10.1016/j.clinph.2014.08.005
11. Ragazzoni A, Pirulli C, Veniero D et al (2013) Vegetative versus minimally conscious states: a study using TMS-EEG, sensory and event-related potentials. PLoS One 8, e57069. doi:10.1371/journal.pone.0057069
12. Wennervirta JE, Ermes MJ, Tiainen SM et al (2009) Hypothermia-treated cardiac arrest patients with good neurological outcome differ early in quantitative variables of EEG suppression and epileptiform activity. Crit Care Med 37:2427–2435. doi:10.1097/CCM.0b013e3181a0ff84

13. Rundgren M, Rosén I, Friberg H (2006) Amplitude-integrated EEG (aEEG) predicts outcome after cardiac arrest and induced hypothermia. Intensive Care Med 32:836–842. doi:10.1007/s00134-006-0178-6
14. Rundgren M, Westhall E, Cronberg T et al (2010) Continuous amplitude-integrated electroencephalogram predicts outcome in hypothermia-treated cardiac arrest patients. Crit Care Med 38:1838–1844. doi:10.1097/CCM.0b013e3181eaa1e7
15. Noirhomme Q, Lehembre R, Zdel RL et al (2014) Automated analysis of background EEG and reactivity during therapeutic hypothermia in comatose patients after cardiac arrest. Clin EEG Neurosci 45:6–13. doi:10.1177/1550059413509616
16. Zandbergen EG, de Haan RJ, Stoutenbeek CP et al (1998) Systematic review of early prediction of poor outcome in anoxicischaemic coma. Lancet 352:1808–1812. doi:10.1016/S0140-6736(98)04076-8
17. Young GB, Doig G, Ragazzoni A (2005) Anoxic-ischemic encephalopathy: clinical and Electrophysiological Associations with outcome. Neurocrit Care 2:159–164. doi:10.1385/NCC:2:2:159
18. Cruse D, Norton L, Gofton T et al (2014) Positive prognostication from median-nerve somatosensory evoked cortical potentials. Neurocrit Care 21:238–244. doi:10.1007/s12028-014-9982-y
19. Wijdicks EF, Hijdra A, Young GB et al (2006) Practice parameter: prediction of outcome in comatose survivors after cardiopulmonary resuscitation (an evidence-based review): report of the Quality Standards Subcommittee of the American Academy of Neurology. Neurology 67:203–210. doi:10.1212/01.wnl.0000227183.21314.cd
20. Jones SJ (2000) Auditory evoked potentials to spectro-temporal modulation of complex tones in normal subjects and patients with severe brain injury. Brain 123:1007–1016. doi:10.1093/brain/123.5.1007
21. Kotchoubey B, Lang S, Baales R et al (2001) Brain potentials in human patients with extremely severe diffuse brain damage. Neurosci Lett 301:37–40
22. Fischer C, Luaute J, Morlet D (2010) Event-related potentials (MMN and novelty P3) in permanent vegetative or minimally conscious states. Clin Neurophysiol 121:1032–1042. doi:10.1016/j.clinph.2010.02.005
23. Cavinato M, Freo U, Ori C et al (2009) Post-acute P300 predicts recovery of consciousness from traumatic vegetative state. Brain Inj. 23(12):973–980. doi:10.3109/02699050903373493
24. Luaute J, Maucort-Boulch D, Tell L et al (2010) Long-term outcomes of chronic minimally conscious and vegetative states. Neurology 75:246–252.
25. Estraneo A, Moretta P, Loreto V et al (2013) Predictors of recovery of responsiveness in prolonged anoxic vegetative state. Neurology 80:464–470. doi:10.1212/WNL.0b013e31827f0f31
26. Wijnen VJ, Eilander HJ, de Gelder B, van Boxtel GJ (2014) Repeated measurements of the auditory oddball paradigm is related to recovery from the vegetative state. J Clin Neurophysiol 31:65–80. doi:10.1097/01.wnp.0000436894.17749.0c
27. Hildebrandt H, Happe S, Deutschmann A et al (2007) Brain perfusion and VEP reactivity in occipital and parietal areas are associated to recovery from hypoxic vegetative state. J Neurol Sci 260:150–158. doi:10.1016/j.jns.2007.04.035
28. Vogel EK, Luck SJ, Shapiro KL (1998) Electrophysiological evidence for a postperceptual locus of suppression during the attentional blink. J Exp Psychol Hum Percept Perform 24:1656–1674. doi:10.1037/0096-1523.24.6.1656
29. Dehaene S, Sergent C, Changeux JP (2003) A neuronal network model linking subjective reports and objective physiological data during conscious perception. Proc Natl Acad Sci U S A 100:8520–8525. doi:10.1073/pnas.1332574100
30. Lamy D, Salti M, Bar-Haim Y (2009) Neural correlates of subjective awareness and unconscious processing: an ERP study. J Cogn Neurosci 21:1435–1446. doi:10.1016/j.neulet.2003.09.060
31. Verleger R (2010) Markers of awareness? EEG potentials evoked by faint and masked events, with special reference to the "attentional blink". In: Czigler I, Winkler I (eds) Unconscious memory representations in perception: processes and mechanisms in the brain. John Benjamin Publishing Company, Amsterdam, pp 37–70

32. Owen AM, Coleman MR, Boly M et al (2006) Detecting awareness in the vegetative state. Science 313:1402
33. Goldfine AM, Victor JD, Conte MM et al (2011) Determination of awareness in patients with severe brain injury using EEG power spectral analysis. Clin Neurophysiol 122:2157–2168. doi:10.1016/j.clinph.2011.03.022
34. Cruse D, Chennu S, Fernandez-Espejo D et al (2012) Detecting awareness in the vegetative state: electroencephalographic evidence for attempted movements to command. PLoS One 7, e49933. doi:10.1371/journal.pone.0049933
35. Guger C, Edlinger G, Harkam W et al (2003) How many people are able to operate an EEG-based brain-computer interface (BCI)? IEEE Trans Neural Syst Rehabil Eng 11:145–147
36. Cruse D, Chennu S, Chatelle C et al (2011) Bedside detection of awareness in the vegetative state: a cohort study. Lancet 378:2088–2094. doi:10.1016/S0140-6736(11)61224-5
37. Cruse D, Chennu S, Chatelle C et al (2012) Relationship between etiology and covert cognition in the minimally conscious state. Neurology 78:816–822. doi:10.1212/WNL.0b013e318249f764
38. Cruse D, Chennu S, Chatelle C et al (2012) Bedside detection of awareness in the vegetative state - Authors' reply. Lancet 379:1702. doi:10.1016/S0140-6736(12)60715-6
39. Goldfine AM, Bardin JC, Noirhomme Q et al (2013) Reanalysis of "Bedside detection of awareness in the vegetative state: a cohort study". Lancet 381:289–291. doi:10.1016/S0140-6736(13)60125-7
40. Cruse D, Chennu S, Chatelle C et al (2013) Reanalysis of "Bedside detection of awareness in the vegetative state: a cohort study" – Authors' reply. Lancet 381:291–292. doi:10.1016/S0140-6736(13)60126-9
41. Owen AM, Coleman MR, Boly M et al (2007) Response to comments on "Detecting awareness in the vegetative state". Science 315:1221c. doi:10.1126/science.1135583
42. Noirhomme Q, Lesenfants D, Gomez F et al (2014) Biased binomial assessment of cross-validated estimation of classification accuracies illustrated in diagnosis predictions. NeuroImage Clin. doi:10.1016/j.nicl.2014.04.004
43. Cruse D, Gantner I, Soddu A, Owen AM (2014) Lies, damned lies, and diagnoses: estimating the clinical utility of assessments of covert awareness in the vegetative state. Brain Inj 28(9):1197–1201. doi: 10.3109/02699052.2014.920517
44. Peterson A, Cruse D, Naci L et al (2015) Risk, diagnostic error, and the clinical science of consciousness. Neuroimage Clin 7:588–597. doi:10.1016/j.nicl.2015.02.008
45. Gibson RM, Fernández-Espejo D, Gonzalez-Lara LE, Kwan BY, Lee DH, Owen AM, Cruse D (2014) Multiple tasks and neuroimaging modalities increase the likelihood of detecting covert awareness in patients with disorders of consciousness. Front Hum Neurosci. doi:10.3389/fnhum.2014.00950
46. Horki P, Bauernfeind G, Klobassa DS, Pokorny C, Pichler G, Schippinger W, Müller-Putz GR (2014) Detection of mental imagery and attempted movements in patients with disorders of consciousness using EEG. Front Hum Neurosci. doi:10.3389/fnhum.2014.01
47. Cruse D, Thibaut A, Demertzi A et al (2013) Actigraphy assessments of circadian sleep-wake cycles in the vegetative and minimally conscious states. BMC Med 11:18. doi:10.1097/00001199-200501000-00005
48. Isono M, Wakabayashi Y, Fujiki MM et al (2002) Sleep cycle in patients in a state of permanent unconsciousness. Brain Inj 16:705–712. doi:10.1080/02699050210127303
49. Landsness EC, Bruno M-A, Noirhomme Q et al (2011) Electrophysiological correlates of behavioural changes in vigilance in vegetative state and minimally conscious state. Brain 134:2222–2232. doi:10.1093/brain/awr152
50. de Biase S, Gigli GL, Lorenzut S et al (2014) The importance of polysomnography in the evaluation of prolonged disorders of consciousness: sleep recordings more adequately correlate than stimulus-related evoked potentials with patients' clinical status. Sleep Med 15:393–400. doi:10.1016/j.sleep.2013.09.026
51. Schiff ND (2012) Moving toward a generalizable application of central thalamic deep brain stimulation for support of forebrain arousal regulation in the severely injured brain. Ann N Y Acad Sci 1265:56–68. doi:10.1111/j.1749-6632.2012.06712.x

52. Espejo DF, Soddu A, Cruse D et al (2012) A role for the default mode network in the bases of disorders of consciousness. Ann Neurol 72(3):335–343
53. Tononi G (2004) An information integration theory of consciousness. BMC Neurosci 5:42. doi:10.1186/1471-2202-5-42
54. Rosanova M, Gosseries O, Casarotto S et al (2012) Recovery of cortical effective connectivity and recovery of consciousness in vegetative patients. Brain 135:1308–1320. doi:10.1093/brain/awr340
55. Ferrarelli F, Massimini M, Sarasso S et al (2010) Breakdown in cortical effective connectivity during midazolam-induced loss of consciousness. Proc Natl Acad Sci U S A 107:2681–2686. doi:10.1073/pnas.0913008107
56. Massimini M, Ferrarelli F, Murphy MJ et al (2010) Cortical reactivity and effective connectivity during REM sleep in humans. Cogn Neurosci 1:176–183. doi:10.1080/17588921003731578
57. Casali AG, Gosseries O, Rosanova M (2013) A theoretically based index of consciousness independent of sensory processing and behavior. Sci Transl Med 5(198):198ra105
58. Azila Noh N, Fuggetta G (2011) Human cortical theta reactivity to high-frequency repetitive transcranial magnetic stimulation. Hum Brain Mapp 33:2224–2237. doi:10.1002/hbm.21355
59. Fuggetta G, Pavone EF, Fiaschi A, Manganotti P (2008) Acute modulation of cortical oscillatory activities during short trains of high-frequency repetitive transcranial magnetic stimulation of the human motor cortex: a combined EEG and TMS study. Hum Brain Mapp 29:1–13. doi:10.1002/hbm.20371
60. Calvo-Merino B, Haggard P (2004) Transcranial magnetic stimulation. Applications in cognitive neuroscience. Rev Neurol 38:374–380
61. Nitsche MA, Seeber A, Frommann K et al (2005) Modulating parameters of excitability during and after transcranial direct current stimulation of the human motor cortex. J Physiol 568:291–303. doi:10.1113/jphysiol.2005.092429
62. Louise-Bender Pape T, Rosenow J, Lewis G et al (2009) Repetitive transcranial magnetic stimulation-associated neurobehavioral gains during coma recovery. Brain Stimul 2:22–35. doi:10.1016/j.brs.2008.09.004
63. Strafella AP, Paus T, Fraraccio M, Dagher A (2003) Striatal dopamine release induced by repetitive transcranial magnetic stimulation of the human motor cortex. Brain 126(Pt 12):2609–2615
64. King JR, Sitt JD, Faugeras F et al (2013) Information sharing in the brain indexes consciousness in noncommunicative patients. Curr Biol 23:1914–1919. doi:10.1016/j.cub.2013.07.075
65. Chennu S, Finoia P, Kamau E et al (2014) Spectral signatures of reorganised brain networks in disorders of consciousness. PLoS Comput Biol 10(10):e1003887. doi:10.1371/journal.pcbi.1003887
66. Piccione F, Cavinato M, Manganotti P et al (2010) Behavioral and neurophysiological effects of repetitive transcranial magnetic stimulation on the minimally conscious state: a case study. Neurorehabil Neural Repair 25:98–102. doi:10.1177/1545968310369802
67. Gabriel D, Henriques J, Comte A et al (2015) Substitute or complement? Defining the relative place of EEG and fMRI in the detection of voluntary brain reactions. Neuroscience 290:435–444. doi:10.1016/j.neuroscience.2015.01.053
68. Kotchoubey B, Lang S, Mezger G et al (2005) Information processing in severe disorders of consciousness: vegetative state and minimally conscious state. Clin Neurophysiol 116:2441–2453
69. Perrin F, Schnakers C, Schabus M et al (2006) Brain response to one's own name in vegetative state, minimally conscious state, and locked-in syndrome. Arch Neurol 63:562–569
70. Schnakers C, Perrin F, Schabus M et al (2008) Voluntary brain processing in disorders of consciousness. Neurology 71:1614–1620
71. Bekinschtein TA, Dehaene S, Rohaut B et al (2009) Neural signature of the conscious processing of auditory regularities. Proc Natl Acad Sci U S A 106:1672–1677
72. Cavinato M, Volpato C, Silvoni S et al (2011) Event-related brain potential modulation in patients with severe brain damage. Clin Neurophysiol 122:719–724. doi:10.1016/j.clinph.2010.08.024

73. Boly M, Garrido MI, Gosseries O et al (2011) Preserved feedforward but impaired top-down processes in the vegetative state. Science 332:858–862. doi:10.1126/science.1202043
74. Höller Y, Bergmann J, Kronbichler M et al (2011) Preserved oscillatory response but lack of mismatch negativity in patients with disorders of consciousness. Clin Neurophysiol 122: 1744–1754. doi:10.1016/j.clinph.2011.02.009
75. faugeras F, Rohaut B, Weiss N et al (2012) Event related potentials elicited by violations of auditory regularities in patients with impaired consciousness. Neuropsychologia 50:403–418. doi:10.1016/j.neuropsychologia.2011.12.015
76. Chennu S, Finoia P, Kamau E et al (2013) Dissociable endogenous and exogenous attention in disorders of consciousness. Neuroimage Clin 3:450–461. doi:10.1016/j.nicl.2013.10.008
77. Steppacher I, Eickhoff S, Jordanov T et al (2013) N400 predicts recovery from disorders of consciousness. Annals of Neurol. doi:10.1002/ana.23835
78. Rohaut B, Faugeras F, Chausson N et al (2015) Probing ERP correlates of verbal semantic processing in patients with impaired consciousness. Neuropsychologia 66:279–292. doi: 10.1016/j.neuropsychologia.2014.10.014
79. Wijnen VJM, van Boxtel GJM, Eilander HJ, de Gelder B (2007) Mismatch negativity predicts recovery from the vegetative state. Clin Neurophysiol 118:597–605. doi:10.1016/j.clinph.2006.11.020
80. Qin PM, Di HB, Yan X et al (2008) Mismatch negativity to the patient's own name in chronic disorders of consciousness. Neurosci Lett 448:24–28
81. Wijnen VJM, Eilander HJ, de Gelder B, van Boxtel GJM (2014) Visual processing during recovery from vegetative state to consciousness: Comparing behavioral indices to brain responses. Neurophysiol Clin 44:457–469. doi:10.1016/j.neucli.2014.08.008

Chapter 10
The Persistent Vegetative State: Evidence That the Lower Brain Survives Because Its Neurons Intrinsically Resist Ischemia

R. David Andrew

Abstract Global ischemia caused by heart attack, pulmonary failure, near-drowning or traumatic brain injury often damages the higher brain but not the brainstem, leading to a 'persistent vegetative state' where the patient is awake but not aware. Approximately 30,000 US patients are held captive in this condition but not a single research study has addressed how the lower brain is preferentially protected in these people. In the higher brain, ischemia elicits a profound anoxic depolarization (AD) causing neuronal dysfunction and vasoconstriction within minutes. Might brainstem nuclei generate less damaging AD and so be more resilient? Here we compared neuronal resistance to acute injury induced by simulated ischemia in 'higher' brain (neocortex, hippocampus, striatum, thalamus and cerebellar cortex) versus 'lower' brain (hypothalamus and brainstem) in live slices from rat and mouse. Light transmittance (LT) imaging in response to 10 min of oxygen/glucose deprivation (OGD) revealed immediate and acutely damaging AD propagating through gray matter of higher regions. In adjacent lower brain nuclei, OGD-evoked AD caused little tissue injury. Whole-cell patch recordings from higher neurons under OGD revealed an immediate and irreversible loss of membrane potential (strong AD) that did not recover. In contrast lower neurons only slowly depolarized (weak AD) and then repolarized post-OGD. Two-photon microscopy confirmed non-recoverable swelling of cortical pyramidal neurons during OGD, while lower neurons appeared uninjured. All of the above responses were mimicked by bath exposure to 100 μM ouabain which inhibits the Na^+/K^+ pump or to 1–10 nM palytoxin which converts the pump into an open cationic channel. Our working hypothesis is that the Na^+/K^+ pump isoforms expressed in lower brain neurons confer resilience during ischemic stress and that higher brain 'shutdown' has evolutionary advantages.

Our studies show that, independent of blood supply, the Na^+/K^+ pump of higher neurons fails quickly and extensively during ischemia compared to naturally

R.D. Andrew
Department of Biomedical and Molecular Sciences, Queen's University, Kingston, ON K7L-3N6, Canada
e-mail: andrewd@queensu.ca

M.M. Monti, W.G. Sannita (eds.), *Brain Function and Responsiveness in Disorders of Consciousness*, DOI 10.1007/978-3-319-21425-2_10

resilient hypothalamic and brainstem neurons. The selective survival of lower brain regions in patients who endure global ischemia will support a persistent vegetative state.

Keywords Stroke • Pyramidal • Hypothalamus • Whole-cell recording • Neuroprotection • TBI • Spreading depolarization • Spreading depression • Anoxic depolarization • Na^+/K^+ pump • ATPase • Ischemic depolarization

10.1 Introduction

Cardiac or pulmonary arrest, strangulation or near drowning are among a number of scenarios that can leave the patient in a persistent vegetative state (PVS). This dire situation evolves from a hypoxic-ischemic encephalopathy involving the entire brain caused by global ischemia. As well brain swelling secondary to traumatic brain injury is a common cause of global ischemia. It was estimated in 1994 that 10,000–25,000 U.S. adults and 4000–10,000 children were in this state of being "awake but not aware" [1, 2]. It has been unclear why the lower but not higher brain survives despite experiencing the same global metabolic stress. The hypothalamus and brainstem can survive to sustain life [3–5] as demonstrated by brain imaging [6–9], by regional brain metabolism measurements [10] and by animal models of global ischemia [11–13]. Longstreth [14] noted that "Brown-Sequard was one of the first to introduce the concept of a rostral-caudal deterioration of brain function with global ischemia. Simply put, cerebral cortical function is lost before brainstem function". This differential loss was apparent from observing patients with intact brainstem reflexes but without higher function (awareness, communication).

Traumatic brain injury (TBI) cardiac/pulmonary failure or near drowning can cause global ischemia that initially induces coma. Within the ensuing 3 weeks, one of 4 clinical scenarios evolves: (1) the *persistent vegetative state* (PVS), (2) the *minimally conscious state* (MCS), (3) return to consciousness, (4) brain death. Due to the high rates of misdiagnosis, the prevalence of PVS is difficult to assess but is estimated at 100/million in the US (~30,000 people) with one third being children [15, 16]. As with PVS, MCS patients 'awake' but also exhibit periods of awareness and limited communication. Given that the brain is globally deprived of blood flow, what makes the brainstem so resilient? To my knowledge only a series of three papers originating from my laboratory has addressed this question to date. We have examined neuronal susceptibility to acute ischemic injury independent of blood flow.

Within about a minute of the onset of global ischemia [17, 18] or stroke [19–21], synapses initially fail and then a profound 'ischemic' or 'anoxic' depolarization (AD) propagates as a front of neuronal inactivation and swelling across the neocortex. Both the initial synaptic silence and then AD result from failure of the Na^+/K^+ pump. Any recovery requires energy from the already compromised gray matter. This causes acute injury to the neurons within minutes if blood flow does not return. During 10 min of oxygen/glucose deprivation (OGD) in live brain slices, the neocortex [22, 23], hippocampus [24, 25], striatum [13], thalamus [26, 27] and cerebellar cortex [28] each undergo AD that then acutely injures the resident neurons over several minutes. The extent of AD reliably determines ensuing neuronal damage in higher brain [21]. In contrast, 'lower' neurons in hypothalamus generate a weak version of AD in vivo [11] and resist acute injury caused by OGD [27, 29]. Previous work showed that some brainstem nuclei of adult rat do not support strong spreading depolarization [30, 31] unless chemically depolarized [32]. We recently showed that, like hypothalamic neurons [29, 33], five types of brainstem neurons better survive the stress of simulated global ischemia (i.e., OGD) by resisting strong AD and by better recovering compared to higher neurons [34].

Most stroke research has focused on injury to structures in the cerebral hemispheres, particularly neocortex and hippocampus. Pyramidal neurons are the output cells of both regions and have been the prototypical neurons studied in terms of stroke damage both in vivo [17, 20, 35, 36] and in live brain slices [22–25, 37]. However it is not well recognized that there is a decreasing rostro-caudal susceptibility of neurons to global brain ischemia [11, 14, 38, 39]. Is this because of regional differences in blood flow or to the properties of the neurons? We have investigated how well neurons of hypothalamus and brainstem survive simulated ischemia. Specifically we examined AD generation induced by oxygen/glucose deprivation (OGD) in higher and lower brain regions in live slices [29, 33, 34] where the extent of simulated ischemia is the same in all regions.

Based on these studies, here I outline support for the 'rostral-caudal hypothesis' by contrasting OGD responses and survivability between higher and lower neurons. We monitored AD propagation by using (1) light transmittance (LT) imaging; (2) whole-cell patch recording; and (3) two-photon laser scanning microscopy in coronal brain slices of adult rodent exposed to 10 min of simulated ischemia. The use of slices rules out that variable regional blood flow influences outcome. We show that unlike higher brain neurons, hypothalamic and brainstem neurons have the intrinsic ability to resist shutdown by AD. By maintaining resting potential, avoiding spike inactivation and rapidly recovering from low-oxygen/low-glucose, brainstem circuitry is better preserved. We propose that this sustains vital bodily functions even as higher brain regions deteriorate. A similar regional selectivity of acute injury to neurons in patients suffering global ischemia can evolve to the persistent vegetative state.

10.2 Materials and Methods

10.2.1 Brain Slice Preparation

All animal care and use was conducted in accordance with the national standards set down by the Canadian Council for Animal Care. Sprague-Dawley rats (21–60 days old; Charles River, St. Constant, PQ) or Wistar-YFP mice (40–100 days old; bred in-house) were decapitated by guillotine under a protocol which included housing standards that was approved by the Queen's University Animal Care Committee. Following craniotomy, the olfactory bulbs and optic tracts were cut. The brain was quickly removed and immersed in ice-cold and oxygenated (95 % O_2, 5 % CO_2) artificial cerebral spinal fluid (aCSF) composed of (in mM) 240 sucrose, 3.3 KCl, 26 $NaHCO_3$, 1.3 $MgSO_4.7H_2O$, 1.23 NaH_2PO_4, 11 D-glucose and 1.8 $CaCl_2$. Using a Leica 1000-T vibratome, 400 μm slices were cut in the sucrose aCSF, usually in the coronal plane. Slices were incubated in regular aCSF (equimolar NaCl replacing sucrose above) at 35 °C for at least 1 h. Slices were then transferred to a recording/imaging chamber where they were submerged in flowing aCSF (3 ml/min.) at 36 °C±0.5 °C. The aCSF osmolality was 295 mOsm at pH 7.4. The oxygen glucose deprivation (OGD) aCSF was of similar composition to control aCSF, except for substituting of 95 % O_2/5 % CO_2 bubbling of aCSF with 95 % N_2/5 % CO_2. In addition, 11 mM glucose was reduced either to 0 mM or 1 mM glucose with osmotic adjustment using NaCl. Following exposure to 10 min of OGD or 100 μM ouabain (Aldrich-Sigma, St. Louis, MO.), slices were reintroduced to control aCSF for 30 min to observe the extent of recovery.

10.2.2 Imaging Changes in Light Transmittance (ΔLT)

Slices were imaged using wide-band illumination through an upright microscope (Axoscope 2FS, Zeiss). Video images were captured with a cooled charged coupled device (Hamamatsu C4742) using Imaging Workbench 6 software (Indec Biosystems Inc.). Each image of a video series consisted of 16 averaged frames acquired at 20 Hz. The first image of the series was the control transmittance (T_{CONT}) which was subtracted from each of the subsequent images (T_{EXP}) in the series. The difference signal was normalized by dividing by T_{CONT}, which varies across the slice depending on the zone sampled. For example, T_{CONT} was lower in white matter than gray matter. This value was then presented as a percentage of the digital intensity of the control image of that series. That is, $\Delta LT = [(T_{EXP} - T_{CONT}) / T_{CONT}] \times 100 = [\Delta T/T]$ %. The change in LT was displayed using a pseudocolor intensity scale. The slice image in bright field was displayed using a gray intensity scale. Note that the initial swelling response to OGD as imaged as an elevated LT response involves both neurons *and* astrocytes while the subsequently reduced LT response indicates injured (beaded) dendrites. Astrocytes quickly recover and do not contribute to the reduced signal [37, 40].

10.2.3 Electrophysiology

Micropipettes were pulled from borosilicate glass (outside diameter 1.2 mm, inside diameter 0.68 mm; World Precision Instruments) to a resistance of 3–6 MΩ. The internal pipette solution contained (in mM) 125 K- gluconate, 10 KCl, 2 $MgCl_2$, 5.5 EGTA, 10 HEPES, 2 Na-ATP and 0.1 $CaCl_2$ (pH was adjusted to 7.3 with KOH).

A 14 mV junction potential was corrected prior to achieving whole-cell configuration. Visually guided whole-cell patch recordings were obtained from pyramidal neurons in neocortical gray or hippocampal CA1 pyramidale, striatum, thalamus, hypothalamus, midbrain-pons (mesencephalic nucleus, MES; locus coeruleus, LC) or in dorsal nucleus solitarius (dNTS) of medulla oblongata and dorsal motor nucleus (DMV) of medulla oblongata. All recordings were acquired in current clamp mode of an Axoclamp 2A amplifier and a Digidata 1322 A/D converter (Molecular Devices). Clampex 10 software (Molecular Devices) was used for data acquisition with subsequent analysis using Clampfit 10 software. Sampling frequency was 10 kHz and low pass filtering was with an external Bessel filter (LPF 202a; Molecular Devices) at 2 kHz. Single neurons were visualized using near-infrared illumination and Dodt gradient contrast optics (Luigs and Neumann, Ratingen, Germany). The class of neuron was identified based (1) on the recorded region in the brain slice which was easily determined from landmarks; (2) on the shape and size of the patched neuron; and (3) on its electrophysiological properties as previously reported by others. Occasionally, lower neurons were exposed to multiple applications of OGD. Also post-OGD, newly acquired recordings could be obtained in lower (but not higher) brain slices.

10.2.4 Two-Photon Laser Scanning Microscopy

Coronal brain slices (400 μm thick) were taken from 30+ day-old C57 black mice of the B6.Cg-Tg (Thy1-YFP)16Jrs/J strain and were prepared as described for rat slices above. The mouse aCSF composition was the same but with 20 mM mannitol added to balance the higher mouse plasma osmolality. These mice have a proportion of pyramidal neurons that express yellow fluorescent protein (YFP) [41] and so the neurons can be imaged in real time during OGD [22, 23, 27, 29, 42, 43]. Structural responses to OGD by mouse slices [37, 43] are similar to the cell swelling and dendritic beading observed in rat slices [44]. An imaging chamber was mounted on a fixed stage of an upright Axioscope II FS microscope (Carl Zeiss, Jena, Germany). YFP^+ neurons were imaged with appropriate filter sets using a Zeiss LSM 710 NLO meta multiphoton system coupled to a Coherent Ti: Sapphire laser. Three-dimensional image stacks were taken at 3.0-μm increments using a Zeiss 40× or 63× water-immersion objective. Data acquisition and analyses were controlled by Zeiss LSM software.

10.2.5 Statistical Analysis

Light transmittance data from each brain region of interest (~500 μm^2) were averaged and plotted +/− standard error. Means were then compared with an independent samples *t*-test, with compared distributions first checked for normality using a one-sample Kolmogorov-Smirnov test, all p's >0.05. Whole-cell patched neurons were analyzed if they displayed stable resting membrane potentials and if the series resistance could be sufficiently compensated. Electrophysiological data were presented as means +/– standard deviation.

10.3 Results

10.3.1 Light Transmittance Imaging

Imaging elevated LT in brain slices or light reflectance (LR) from the neocortical brain surface reveals cell swelling-associated signals (LT increases or LR decreases) as the front of LT change propagates across gray matter [17, 43]. In the wake of this front, which has been well documented as representing encroaching AD, synaptic spines disappear as dendritic beads form. The beading scatters light thereby decreasing LT and increasing LR. Within 6 min of OGD the gray matter of neocortex (Fig. 10.1a), hippocampal formation, striatum, thalamus and cerebellar cortex (Fig. 10.1b) independently display striking AD onset and propagation. Wherever the AD front spreads (Fig. 10.1b, arrows), light scatter denoting injury arises within several minutes (Fig. 10.1a1, a2, purple pseudocoloring). This propagating sequence of swelling and then injury is replicated in slices of higher brain regions by inhibiting the Na^+/K^+ pump with bath exposure to 100 μM ouabain (Fig. 10.1a2) or by converting the pump into an open cationic channel using 1–10 nM palytoxin (not shown) [34]. In neocortex, striatum, hippocampus and thalamus either treatment evokes AD-like LT changes similar to OGD. That is, an elevated LT front of cell swelling denoting AD onset gives rise to an irreversible decrease in LT wherever the front has propagated (Fig. 10.1a2).

Fig. 10.1 (**a**) Anoxic Depolarization injury in necortex. Light transmittance (LT) imaging reveals damage following AD in higher, but not lower, brain regions. (**a1**) The front of LT change representing AD onset arises and propagates as a distinct 'wave', coursing through neocortical gray matter (*arrows*) and leaving damage (light scatter, magenta) in its wake. The white square delineates a region of interest (ROI) of ~30×30 pixels. (**a2**) Light transmittance change in the ROI is plotted over time in response to OGD (red trace). A similar response is evoked by the Na^+/K^+ pump inhibitor ouabain (100 μM) in another neocortical slice. (**b**) AD injures in cerebellar cortex but not brainstem. In a slice containing cerebellum (*Cbell*) and midbrain-pons, AD begins propagating through cerebellar gray after ~3 min of OGD. By 4:30 AD 'mini-fronts' start spreading through each tegmental nucleus (*TN*) in the pons. By 26 min AD propagation has injured cerebellar gray, but not TN or adjacent brainstem regions. The inferior colliculi (*InfCo*) display some superficial swelling and then damage. (**c**) Regional peak gray matter changes in LT evoked by OGD. (**c**) AD generation is stronger and more damaging in rostral brain regions compared to brainstem. (**c1**) Comparison of regional responses to OGD in terms of AD strength (peak LT increases) and subsequent dendritic injury (maximal LT decreases) shows more swelling and subsequent injury in higher brain regions compared to brainstem nuclei. The exception is superior colliculus (*SupCo*) which displays injury similar to higher gray matter, possibly because it has been a prominent surface structure throughout vertebrate evolution until recently. (**c2**) The means of peak LT values displayed by higher and lower regions (excluding *SupCo*) were collapsed and compared with an independent samples *t*-test and found to be significantly different [$t(58)=3.6$, $p<0.001$]

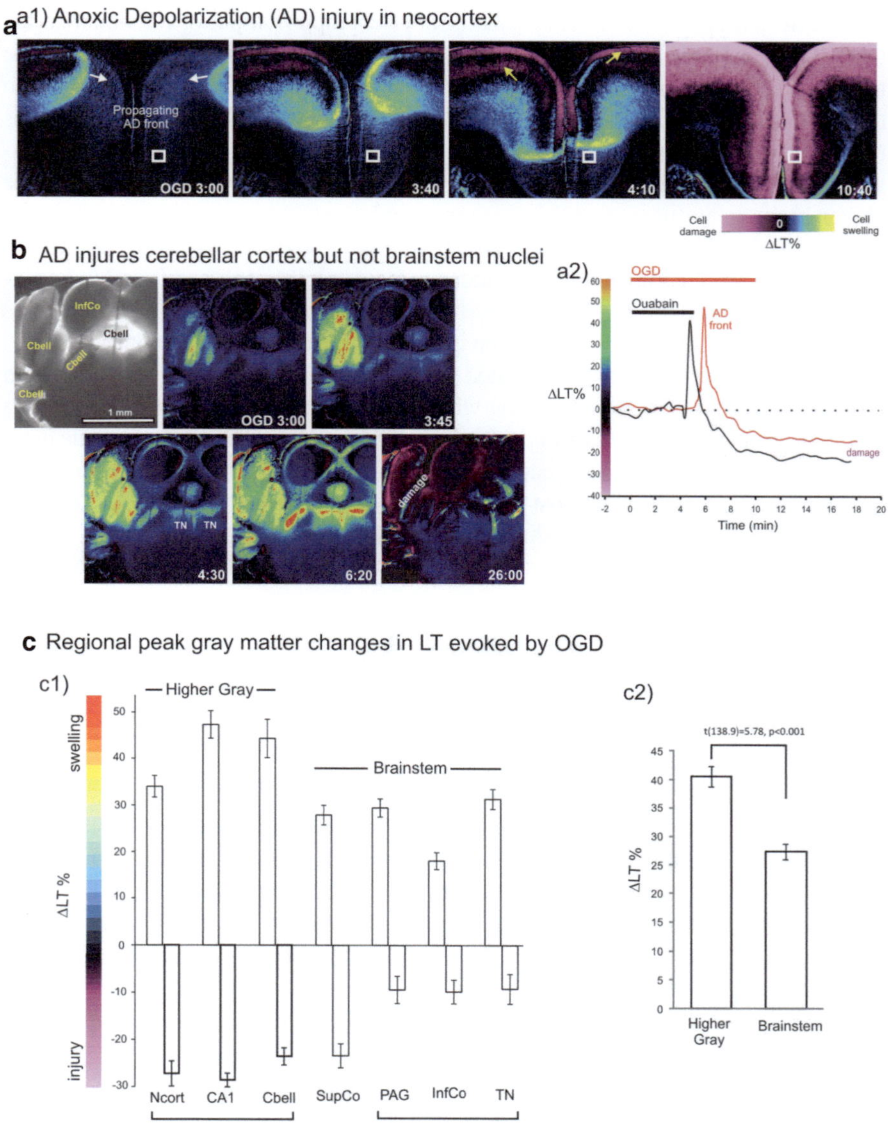
a
a1) Anoxic Depolarization (AD) injury in neocortex
Propagating AD front
OGD 3:00
3:40
4:10
10:40
Cell damage
0
Cell swelling
ΔLT%
b AD injures cerebellar cortex but not brainstem nuclei
InfCo
Cbell
Cbell
Cbell
Cbell
1 mm
OGD 3:00
3:45
TN
TN
4:30
6:20
damage
26:00
a2)
OGD
Ouabain
AD front
ΔLT%
damage
Time (min)
c Regional peak gray matter changes in LT evoked by OGD
c1)
Higher Gray
Brainstem
swelling
ΔLT %
injury
Ncort
CA1
Cbell
SupCo
PAG
InfCo
TN
c2)
t(138.9)=5.78, p<0.001
ΔLT %
Higher Gray
Brainstem

In OGD slices containing brainstem, the adjacent hippocampus or cerebellar cortex (Fig. 10.1b) supports a spreading LT front with subsequent damage. At the same time the brainstem nuclei may display OGD-induced 'mini-fronts' of AD coursing through the periaqueductal gray, superior colliculus (not shown) and tegmental nucleus (Fig. 10.1b, c). However these lower brain regions display little subsequent light scatter with the exception of superior colliculus (Fig. 10.1c). LT values from the three higher brain areas (neocortex, hippocampus and cerebellar cortex) and three lower brain areas (PAG, IC and TG) were collapsed and the means plotted in Fig. 10.1c2. These means were compared with an independent samples *t*-test and found to be significantly different [$t(138.9) = 5.78$, $p < 0.001$]. Similarly the three higher regions showed significantly more LT reduction than the three lower regions (not shown).

These observations which indicated less AD-evoked damage in the brainstem post-OGD were followed up by using whole-cell patch recording of neurons during OGD to compare responses of higher gray matter with lower brain nuclei. The electrophysiological properties of each recorded neuron as well as the properties of lower neurons recorded in post-OGD slices can be found at http://hdl.handle.net/1974/7578 [45]. General findings are presented below.

10.4 Higher Neuron Electrophysiology

10.4.1 Neocortical Pyramidal Cells

Twenty pyramidal neurons (PyNs) were monitored by whole-cell patch recording during simulated stroke (Fig. 10.2a, top trace). The patch pipette was visually placed within layer V and the targeted PyNs identified based on the cell's triangular shape and diameter of 15–20 μm. The resting membrane potential of the 20 cells averaged -82 ± 2.5 mV. Mean input resistance and action potential amplitude were 82 ± 31 MΩ and 79 ± 6.6 mV, respectively (Table 10.1) as typically reported for PyNs in other studies.

Exposing PyNs to 10 min of OGD ($n = 20$) elicited an abrupt anoxic depolarization (AD) with a mean slope of 94.1 ± 40.1 mV/min and a mean onset time of 4.1 ± 0.8 min; (Fig. 10.2a top trace; Table 10.1). The maximal average membrane depolarization during AD was to 0 ± 2.3 mV. Upon return to control aCSF this slowly returned to a mean of 27 % of the original membrane potential. Following a

Fig.10.2 (continued) (**b**) Identical OGD (*left*) as shown in A but by a hypothalamic neuron in the supraoptic nucleus. This magnocellular neuroendocrine cell typically shows a weak AD with spike inactivation. The cell repolarizes with spike reactivation, signaling recovery. (**c**) Two-photon laser scanning microscopy reveals that PyNs are injured after AD. PyN cell bodies and proximal dendrites in cortical layer V dramatically swell as observed 20 min after OGD in the green fluorescent protein (GFP) mouse. The distribution graph (*right*) shows single PyN and MNC responses to OGD. PyNs swell by an average of 34.5 ± 11.0 % in response to 10-min OGD. MNCs shrink slightly by an average of 3.4 ± 4.2 % in response to 15-min OGD. XS area, cross-sectional area

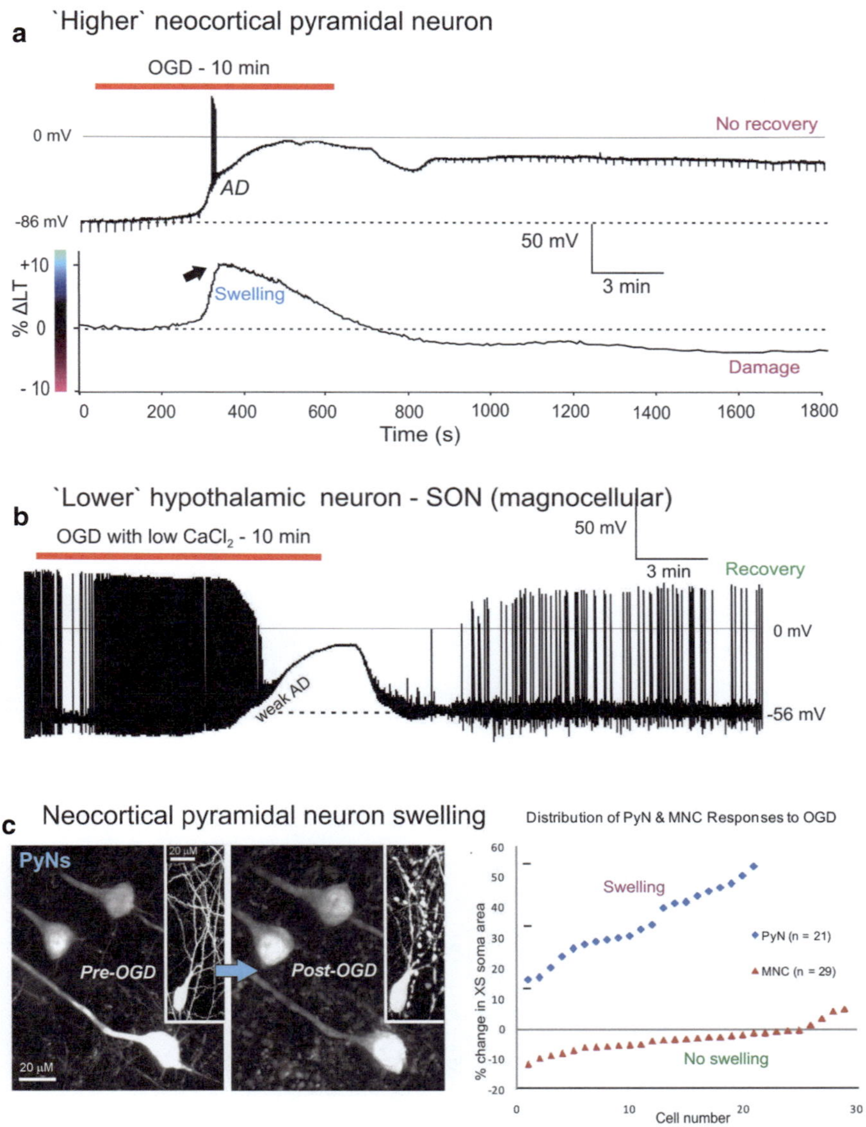

Fig. 10.2 (**a**) 'Higher' neocortical pyramidal neuron. (**b**) 'Lower' hypothalamic neuron – SON (magnocellular). (**c**) Neocortical pyramidal neuron swelling. (**a**) Pyramidal neurons do not recover following OGD. Top trace: Whole-cell PyN recording during 10 min of OGD. Downward deflections represent voltage response to 70 pA hyperpolarizing current pulses elicited every 20 s to monitor whole-cell input resistance. Post-AD this value remains near zero megaohms, confirming injury. Bottom trace: Simultaneous imaging of the cortical field in layer V. Light transmittance change (ΔLT) is plotted over time in a100 μm × 50 μm ROI which includes the PyN recorded above. A 9 % LT increase (*arrow*) corresponds with the fast component of the AD, as the AD wave front reaches the recording pipette tip at ~320 s

Table 10.1 Summary of whole-cell recording data for higher and lower projection neurons in response to OGD

Neuron type	Cell no.	Rmp (mV)	Recovered cell no. Post-OGD	Rmp Recov (%)	Max Depol. (mV) (%)	AP Ampl Recov (%)	Rin Recov	AD Onset (s)	AD Rate (mV/s)	Ref.
Higher gray										
Neocortical PyN	20	−82 ± 2.5	0	27	0	0	57	246	1.6	[29]
CA1 PyN	9	−63 ± 2.5	0	10	0	0	–	343	3.0	[34]
Striatum	10	−70 ± 3.0	0	29	0	0	64	287	3.0	[34]
Thalamus (AMV)	12	−70 ± 3.5	0	–	−4	0	–	217	2.0	[27]
(midline)	7	−69 ± 2.9	0	–	−9	0	–	243	–	[27]
Hypothalamus										
MNC (SON)	21	−65 ± 2.9	11	87	−19	82	84	484	0.2	[29]
MNC (PVN)	8	−51 ± 3.9	7	84	−9	82	89	298	–	[27]
Parvo (PVN)	12	−49 ± 4.2	11	78	−6	95	92	298	–	[27]
SCN	18	−45 ± 4.1	15	81	−19	65	100	431	–	[27]
Brainstem										
Mesenceph. Nuc	14	−53 ± 2.3	7	93	−42	95	86	301	<0.5	[34]
Locus Ceruleus	8	−48 ± 2.8	5	82	−4	83	100	301	<0.5	[34]
Nuc Solitarius	31	−48 ± 4.5	12	75	−10	73	57	322	0.3	[34]
Dorsal Motor Nuc	12	−46 ± 1.4	10	98	−1	73	95	322	0.3	[34]

Four types of higher brain neuron are compared to seven types of lower neurons

Note the poor recovery statistics of higher neurons compared to the lower neuron statistics Standard deviation values not included. Resting membrane potentials were originally adjusted for a + 14 mV junction potential. This adjustment does not affect % recovery of the membrane potential

single 10 min OGD exposure, subsequent intracellular recordings from the same neocortical slice could no longer be obtained, even up to 3 h after return of oxygen and glucose. Importantly it was no longer possible to record evoked field potentials from layers II-III or V post-OGD upon stimulation of layer VI (not shown) so the entire neocortex was electrically silent.

Light transmittance (LT) imaging of the pyramidal cell field during whole-cell recording demonstrated a sudden increase in ΔLT corresponding with AD onset in the single cell (Fig. 10.2a, lower trace). PyNs slowly depolarized during initial OGD, followed by the fast component of the AD. An immediate 8 % increase in ΔLT coincided with the propagating AD wave front, which in this case reached the recorded PyN at ~ 320 s of OGD, coinciding with the fast AD component (Fig. 10.2a, lower trace, arrow). Upon reintroduction of aCSF, membrane potential returned to only 32 % of its original value after 20 min. Action potentials could not be evoked post-OGD because of this maintained depolarization where Na^+ channels were inactivated. There was no subsequent recovery.

10.4.2 Hippocampal CA1 Neurons

The patch pipette was visually placed within layer CA1 and pyramidal somata targeted based on the triangular shape and diameter of 15–20 μm. The average resting membrane potential was −63 ± 2.5 mV (Table 10.1). Mean action potential amplitude and input resistance were 84 ± 13.4 mV and 99 ± 18 MΩ, respectively. Nine CA1 pyramidal neurons were monitored by whole-cell patch during simulated stroke. Ten minutes of OGD elicited an abrupt AD in CA1 neurons that reached zero millivolts in all cells tested (Fig. 10.3a1). Mean AD onset time was 343 ± 85 s. Seven cells responded to OGD with an initial hyperpolarization (*H*, Fig. 10.3a1). Following 20 min recovery in control aCSF the mean membrane potential was only −20 ± 7.4 mV, representing a mere 10 % recovery (Table 10.1). Whole-cell input resistance was too low to accurately measure post-OGD. No new recordings could be acquired in the CA1 stratum pyramidale post-OGD as the neurons were all permanently depolarized. The response to OGD by the nine cells was indistinguishable from 17 other CA1 neurons in our previous study [46] so we did not expand the pool further.

10.4.3 Thalamic Neurons

Eleven neurons in anteromedial ventral (AMV) nucleus were monitored by whole-cell current clamp during simulated stroke (Fig. 10.4a; Table 10.1). The patch pipette was visually placed within AMV and the cells targeted based on their elongated shape and length of 20–30 μm. When steadily hyperpolarized below −65 mV, a depolarizing current step evoked a low threshold potential driving a burst of action potentials (Fig. 10.4a, left), typical of thalamic neurons. The mean resting potential of the 11 AMV cells was −70 ± 3.2 mV. Mean input resistance and action potential amplitude were 81 ± 20 MΩ

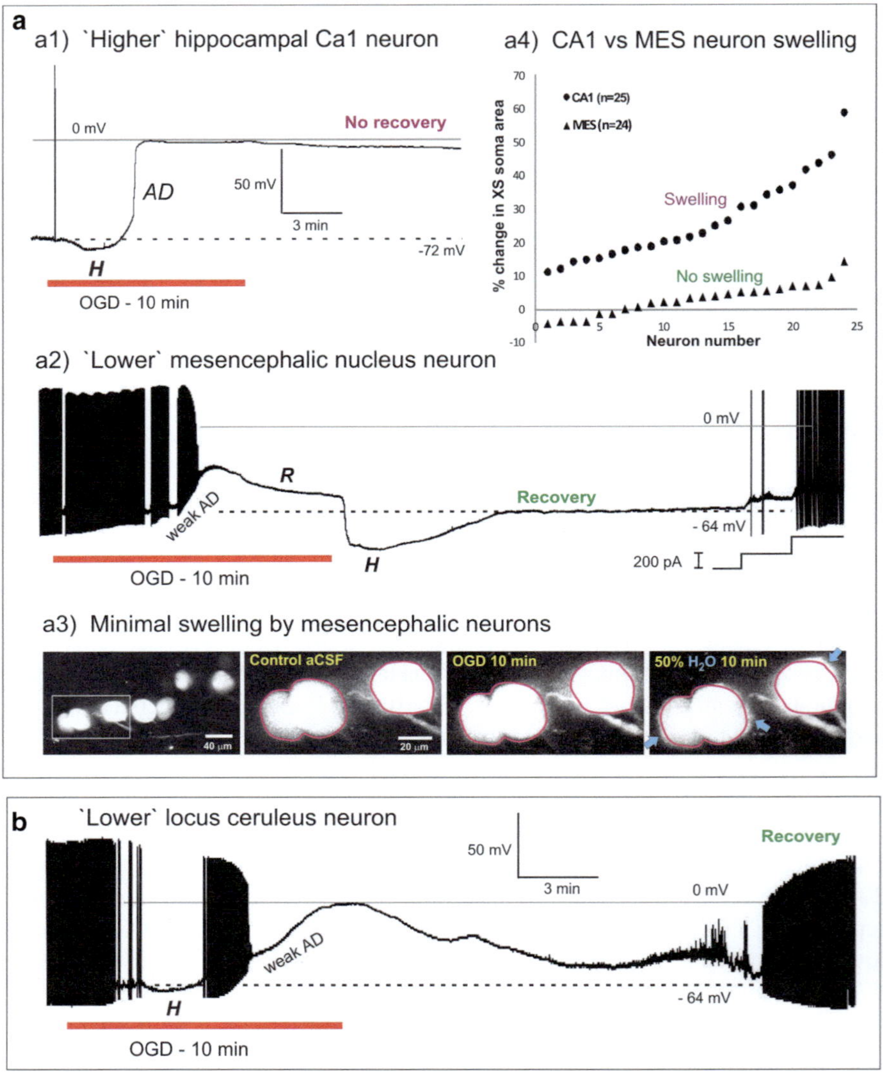

Fig. 10.3 (**a1**) 'Higher' hippocampal CA1 pyramidal neurons undergo terminal AD induced by OGD (as do hippocampal CA3 neurons [76] and granule cells [46]. In response to 10 min of OGD, the CA1 neuron initially hyperpolarizes before undergoing rapid AD to near-zero millivolts from where it does not recover. (**a2**) 'Lower' mesencephalic nucleus neuron. MES neurons resist depolarization and swelling in response to OGD. Whole-cell recording of MES neuron. The spontaneously firing neuron slowly depolarizes and inactivates in response to 10 min of OGD. Surprisingly, the cell then *re*polarizes *during* OGD (*R*). With return of control aCSF, the neuron displays a hyperpolarizing undershoot (*H*), before regaining its original membrane potential. At the end of the recording, steps of depolarizing current demonstrate the neuron's ability to discharge. (**a3**) Minimal swelling by mesencephalic neurons. Volume responses to OGD by YFP-positive MES neurons monitored in real time with 2-photon microscopy. Cell bodies in the MES nucleus resist swelling after 10 min of OGD. In contrast, post-OGD dilution of the aCSF by 50 % with distilled water does evoke swelling (*blue arrows*). (**a4**) CA1 vs MES neuron swelling. Change in cross-sectional (*XS*) area of the cell body is measured to determine soma

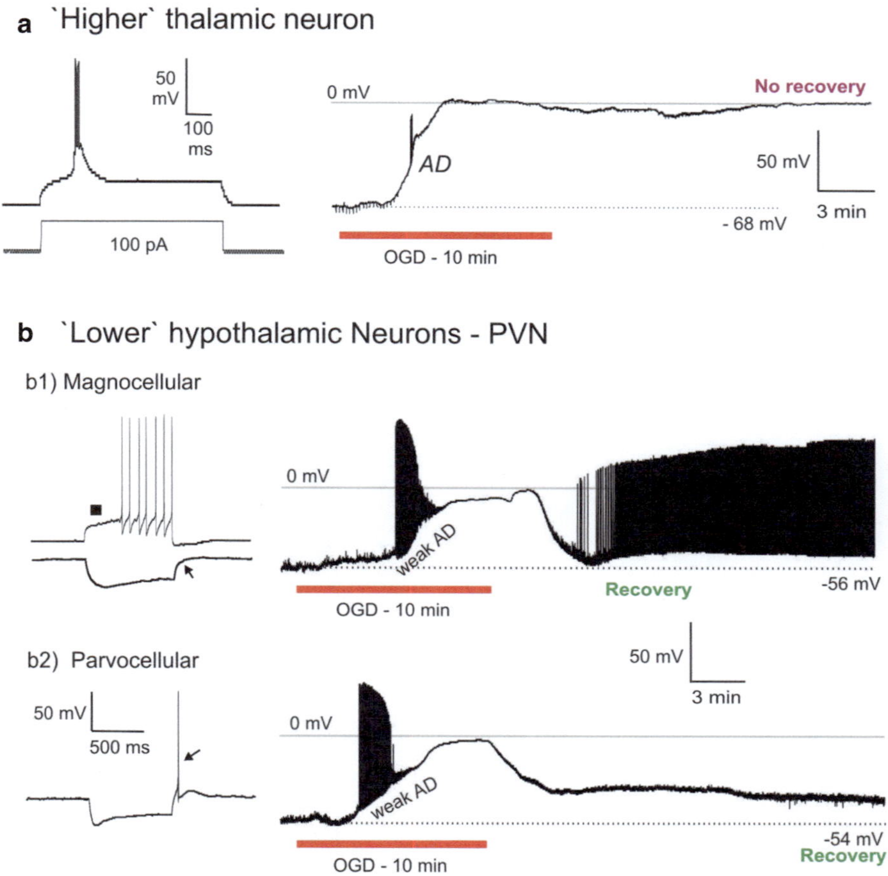

Fig. 10.4 (**a**) 'Higher' thalamic neuron. A thalamic neuron held hyperpolarized by 10 mV (*left*) characteristically fires a brief burst when step depolarized. At a more positive resting potential, the same neuron fires tonically (not shown). During 10 min of OGD (*right*), a thalamic neuron undergoes rapid AD to near-zero millivolts. Following return to control aCSF the cell could not repolarize. This is typical for higher neurons following 10 min of OGD in slices at 34–36 °C. (**b**) 'Lower' hypothalamic nerurons – PVN. (**b1**). Magnocellular. (**b1**) MNCs display a characteristic delay in firing (*square*) when depolarized from negative resting potentials and a delay in return to baseline (*arrow*) in response to a hyperpolarizing current step. During 10 min of OGD (*right*), an MNC gradually depolarizes before reaching a plateau as action potentials inactivate. Following return to control aCSF, the MNC rapidly repolarizes with some action potential firing. (**b2**) A non-bursting parvocellular neuron with characteristic post-inhibitory rebound spike (*arrow*). The neuron responds to 10 min of OGD by gradually depolarizing before reaching a plateau as action potentials inactivate. The neuron resists further depolarization before slowly approaching zero millivolts. Following return to control aCSF, the neuron repolarizes to near-baseline

Fig.10.3 (continued) swelling or shrinking in real time. Swelling by CA1 neurons is pronounced compared to the small range displayed by 'lower' MES neurons in response to 10 min OGD. Results here are similar to Fig. 10.2 comparing neocortical pyramidal neurons and 'lower' hypothalamic magnocellular endocrine neurons. (**b**) 'Lower' locus cerules neuron. LC neurons resist depolarization in response to OGD. Initially, the neuron briefly hyperpolarizes (*H*) as action potentials inactivate before reaching a plateau. It then continues to near-zero millivolts. Upon return to control aCSF, the LC neuron slowly repolarizes

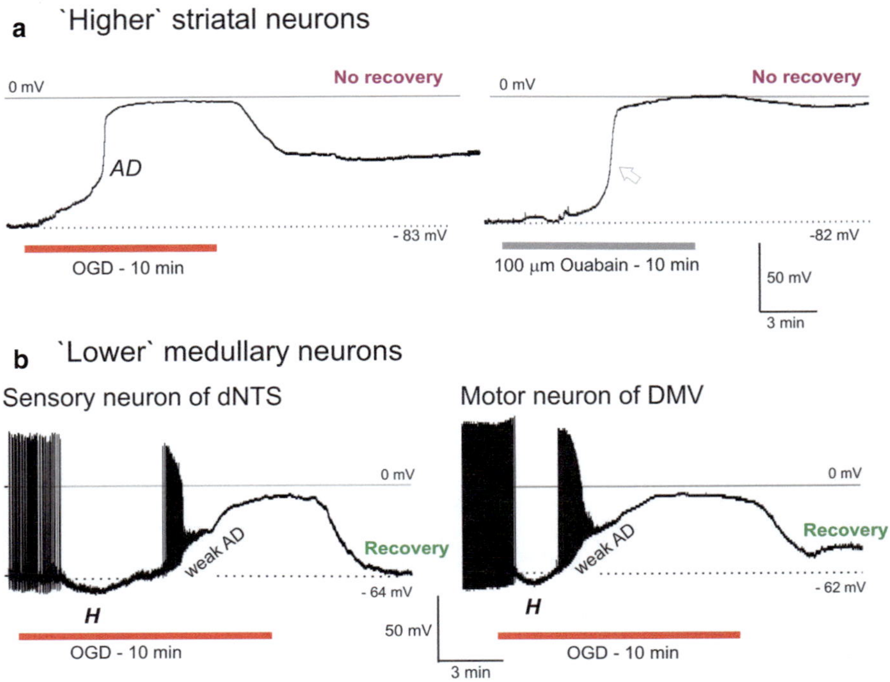

Fig. 10.5 (**a**) In response to OGD (*left*) a striatal neuron slowly depolarizes before undergoing rapid AD to near-zero millivolts. Upon return to control aCSF, the neuron begins to repolarize, but does not recover. In response to the Na^+/K^+ pump blocker ouabain (*right*), the response is similar to OGD. (**b**) Upon exposure to OGD, a sensory neuron and a motor neuron in medulla oblongata initially hyperpolarize (*H*) and then gradually depolarize with eventual spike inactivation, reaching a plateau that is maintained throughout OGD. Upon return of control aCSF each neuron repolarizes to near resting potential

and 78±5.9 mV, respectively. Exposing these 11 neurons to OGD elicited an abrupt anoxic depolarization (AD) with a mean onset time of 217±26 s (Fig. 10.4a, right). The average maximal depolarization reached during AD, was −4±2.5 mV. Average resting membrane potential following OGD could not be determined, as none of the whole-cell recordings from the AMV nucleus could be maintained.

To show that the AMV neurons were typical thalamic cells, additional recordings were made in other thalamic nuclei (n=8). Each of these neurons showed abrupt AD in response to OGD (Table 10.1) as described above. And as in other higher regions, even after an hour of normal aCSF exposure post-AD it was impossible to obtain new recordings from the electrically silent thalamus.

10.4.4 Striatal Neurons

The patch pipette was visually placed within the striatum and cells targeted based on their diameter of 10–15 μm. Spiny projection neurons fire tonically in response to a depolarizing stimulus, typical of striatal projection neurons [13] (not shown).

The average resting membrane potential of the 10 cells was −70 ± 3.0 mV. Input resistance and action potential amplitude were 55 ± 18 MΩ and 81 ± 9.1 mV, respectively, as reported in other studies [47].

Ten striatal neurons were monitored during 10 min of OGD which evoked an abrupt AD to near-zero millivolts (Fig. 10.5a, left). Mean AD onset was 287 ± 23 s. Upon reintroduction of control aCSF, neurons repolarized to −20 ± 16.9 mV, representing only a 29 % recovery with further deterioration following. Cell input resistance did not recover. As with other higher neurons described above, no action potentials could be evoked post-OGD. As well, no newly acquired recordings could be obtained in striatum, further indicating regional injury.

Exposing 4 striatal neurons to 100 μM ouabain for 10 min elicited an abrupt AD-like response (arrow, Fig. 10.5a, right) that approached zero millivolts. As with OGD, striatal neurons fired no action potentials during the depolarization phase and no substantial recovery occurred. In fact although not detailed in this article, ouabain exposure mimicked the OGD response in every neuronal type tested [29, 33, 34]. *This emphasizes the central importance of Na^+/K^+ pump failure in the ischemic response by neurons of every region in the CNS.*

10.5 Lower Neuron Electrophysiology

Our definition of the 'lower' brain is the hypothalamus and brainstem. As described in this chapter and detailed in [33], the thalamus-hypothalamus interface represents a distinct border where more caudal neurons take on an elevated resistance to OGD (shown below) compared to the neurons just described.

10.5.1 Hypothalamic Neurons

Whole-cell data from thalamic ($n = 19$) and hypothalamic neurons ($n = 38$) in response to 10 min OGD are summarized in Table 10.1. As noted above, post-OGD thalamic membrane potentials were near-zero. In sharp contrast, hypothalamic neurons located only a millimeter or less away responded to identical OGD by only slowly depolarizing and usually recovering when back in control aCSF. This was true whether the neurons were recorded in SON (Fig. 10.2b), PVN (Fig. 10.4b) or SCN [33]. As well, there was no immediate spike inactivation as seen in higher neurons. As with all lower brain neurons described below, resistance to OGD-induced depolarization was seen despite a more positive resting membrane potentials pre-OGD (Table 10.1).

Figure 10.2b demonstrates that MNCs are not in some way protected from irreversible depolarization by inhibitory synaptic input. Superfusion of OGD aCSF with calcium chloride reduced from 1.8 to 0.45 mM blocks any residual synaptic input not removed by OGD itself. Yet the depolarization remains intact, indicating that the response is intrinsic to the recorded neuron.

10.5.2 Midbrain/Pontine Neurons

Neurons in Locus Ceruleus (LC) Eight neurons from the LC were whole-cell patched. The pipette was visually placed within the LC and the cells targeted based on their elongated shape and diameter of 20 μm. LC neurons were further identified by their electrophysiological properties [34]. Ten minutes of OGD only gradually depolarized LC neurons, reaching −28 ± 3.5 mV (Fig. 10.3b). Spike inactivation also developed slowly. After this plateau, LC neurons resisted further depolarization for an average duration of 57 ± 21 s, before further depolarizing to −4 ± 4.6 mV. The average membrane potential of 6 LC neurons that were held following 10 min of OGD was −40 ± 10.1 mV, representing 87 % recovery (Table 10.1). The action potential amplitude recovered by 83 %.

Neurons in Trigeminal Mesencephalic (MES) Nucleus A patch pipette was visually placed within MES nucleus and the cells targeted based on their large diameter of 25–50 μm and their round or ellipsoid shape [48]. MES neurons were further identified by their electrophysiological properties [34]. Exposing 14 MES neurons to 10 min of OGD elicited only slight depolarization. The majority were silent but some initially fired spontaneously prior to OGD (Fig. 10.3a2). Of note, all MES neurons actively *re*polarized *during* OGD (*R*, Fig. 10.3a2). Upon return to control aCSF, MES neurons often displayed a hyperpolarizing undershoot (*H*, Fig. 10.3a2) before returning to baseline levels. The average resting membrane potential of 12 MES neurons that could be held following 10 min of OGD displayed a 93 % recovery (Table 10.1). Input resistance recovered by 86 % and action potential amplitude 95 %. Moreover healthy MES neurons (n = 7) could be readily patched in slices previously exposed to 10 min of OGD. Their average resting membrane potential represented a 99 % recovery compared to MES neurons not previously exposed to OGD. Input resistance recovered by 100 % and action potential recovered by 97 %. Some MES neurons were subjected to additional 10 min OGD exposures with near-complete recovery [34]. Exposing three MES neurons to 100 μM ouabain for 10 min caused a gradual depolarization, similar to OGD and gradual recovery [34]. As with OGD, the recorded neuron resisted further depolarization during ouabain exposure, reaching a maximal depolarization of −39 mV before gradually recovering in aCSF.

10.5.3 Medullary Neurons

Neurons in Dorsal Nucleus Tractus Solitarius (dNTS) The pattern of a slow AD induced by OGD followed by recovery continued in the brainstem. Thirty-one sensory neurons from dNTS of the medulla oblongata were whole-cell patched during simulated stroke. The patch pipette was visually placed within the dNTS and the cells targeted based on their diameter of 15–20 μm. They were further categorized based on established electrophysiological properties [34]. Exposing dNTS neurons to 10 min (n = 15) or 15 min

($n=16$) of OGD, elicited a gradual depolarization ($n=7$, Fig. 10.5b) often preceded a brief initial hyperpolarization ($n=24$, Fig. 10.5b) as in LC neurons (Fig. 10.3b). The membrane potential of 17 dNTS neurons that were still held on to following 10 or 15 min of OGD displayed a 75 % recovery. Input resistance and action potential amplitude displayed a 57 % and 73 % recovery, respectively. Healthy neurons from the dNTS could be readily patched *following* 10 or 15 min of OGD. The average resting membrane potential of 12 dNTS neurons that were newly acquired showed a 95 % recovery compared to neurons *not* previously exposed to OGD. Mean cell input resistance displayed an 89 % recovery and action potential amplitude recovery was 80 %.

Neurons in Dorsal Motor Vagal (DMV) Nucleus The patch pipette was visually placed within the DMV nucleus and the cells targeted based on their ellipsoid shape and diameter of ~20 μm [49]. DMV neurons were further identified by their electrophysiological properties [34]. Twelve neurons from the DMV nucleus were recorded during simulated stroke. OGD elicited either a gradual depolarization ($n=5$, not shown) or a brief hyperpolarization ($n=7$, Fig. 10.5b) accompanied by slow spike inactivation followed by a depolarizing plateau. With return to control aCSF, DMV neurons repolarized (Fig. 10.5b) although most recordings were difficult to maintain. However newly acquired DMV neurons could be readily recorded in slices recovered from 10 min OGD. The average resting potential of 10 such DMV neurons showed a 98 % recovery compared to neurons not previously exposed to OGD. Mean input resistance recovered by 95 % and action potential amplitude by 73 %. Clearly DMV neurons could nicely recover.

Characteristic responses to OGD by the neuronal types examined above are summarized in Figs. 10.6 and 10.7. Higher neurons rapidly spike-inactivated and then completely depolarized during OGD and could not recover (Fig. 10.6). Hypothalamic and brainstem neurons displayed less polarized resting membrane potentials (Table 10.1) yet better resisted OGD, responding with a delayed and weak depolarization, followed by repolarization and recovery post-OGD (Fig. 10.7).

10.6 Two-Photon Microscopy

We carried out two separate microscopical studies of live neurons in brain slices to compare higher and lower neuronal responses to OGD independent of whole-cell recordings. It is well documented that OGD induces neuronal soma swelling within a minute or so of AD initiation [37]. In our first study [29], fluorescent rat neocortical pyramidal cells were imaged and changes in their cross-sectional areas post-OGD were measured. These values were contrasted with hypothalamic MNC values under the same conditions measured in the supraoptic nucleus. Plotting each neuron's change in area shows that neocortical pyramidal cells swell acutely and significantly while the hypothalamic neurons do not shift in volume when similarly stressed (Fig. 10.2c, left). The same approach was taken using hippocampal pyramidal neurons, comparing their volume response to midbrain-pontine neurons in the mesencephalic nucleus [34]. Again the higher pyramidal cell bodies swell whereas the midbrain cells do not increase in size despite OGD for up

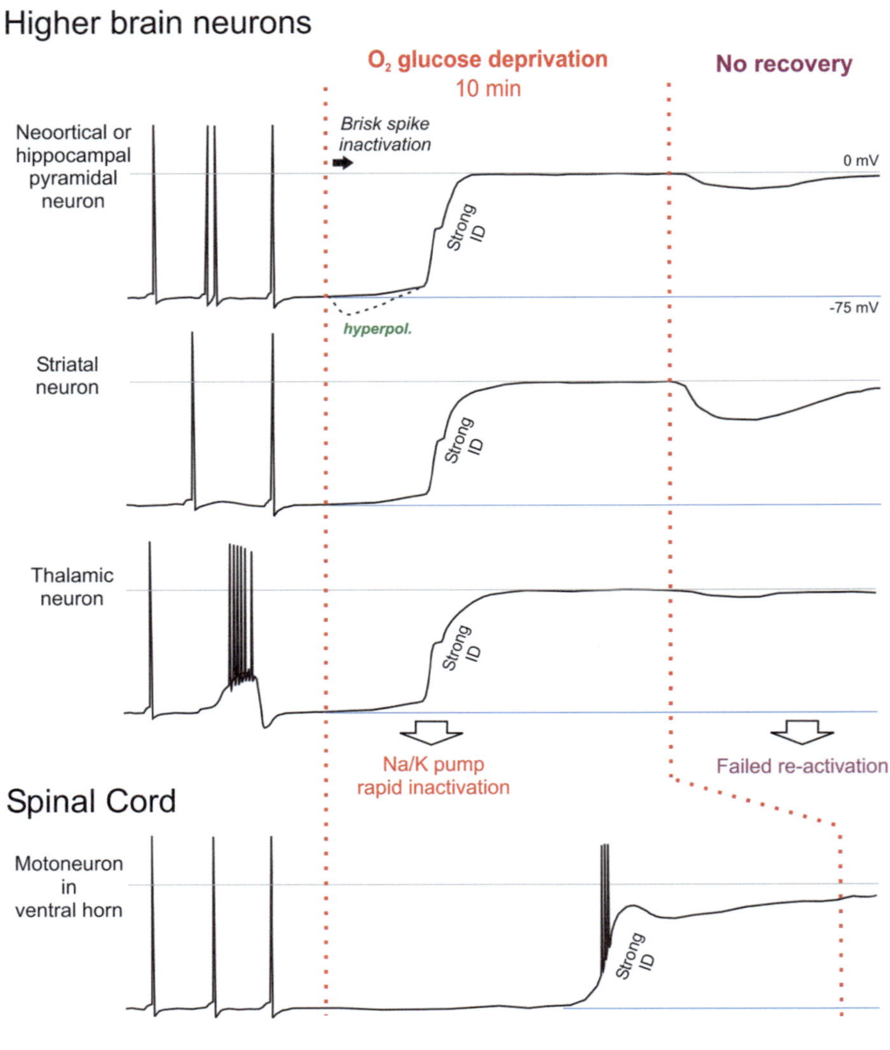

Fig. 10.6 Summary of higher neuron responses to simulated ischemia. Pyramidal, striatal and thalamic neurons immediately spike-inactivate and pyramidal neurons may initially and transiently hyperpolarize. A slow depolarization then rapidly evolves to a pronounced AD, here labelled ID (ischemic depolarization) which quickly approaches zero millivolts. In slices, these higher neurons do not recover from OGD once AD occurs. A similar fate awaits motoneurons recorded in slices of spinal cord under OGD [60]

to 15 min (Fig. 10.3a3, a4). To show that swelling is detectable in these cells, we briefly diluted the aCSF by 50 % with distilled water. Although mammalian neurons generally lack water channels (aquaporins) [37], water will diffuse inward under such a large gradient and indeed the MES neurons swell as expected (Fig. 10.3a4, right). This supports the idea that OGD does not induce the level of metabolic stress in lower neurons that leads to irreversible depolarization and swelling of higher neurons.

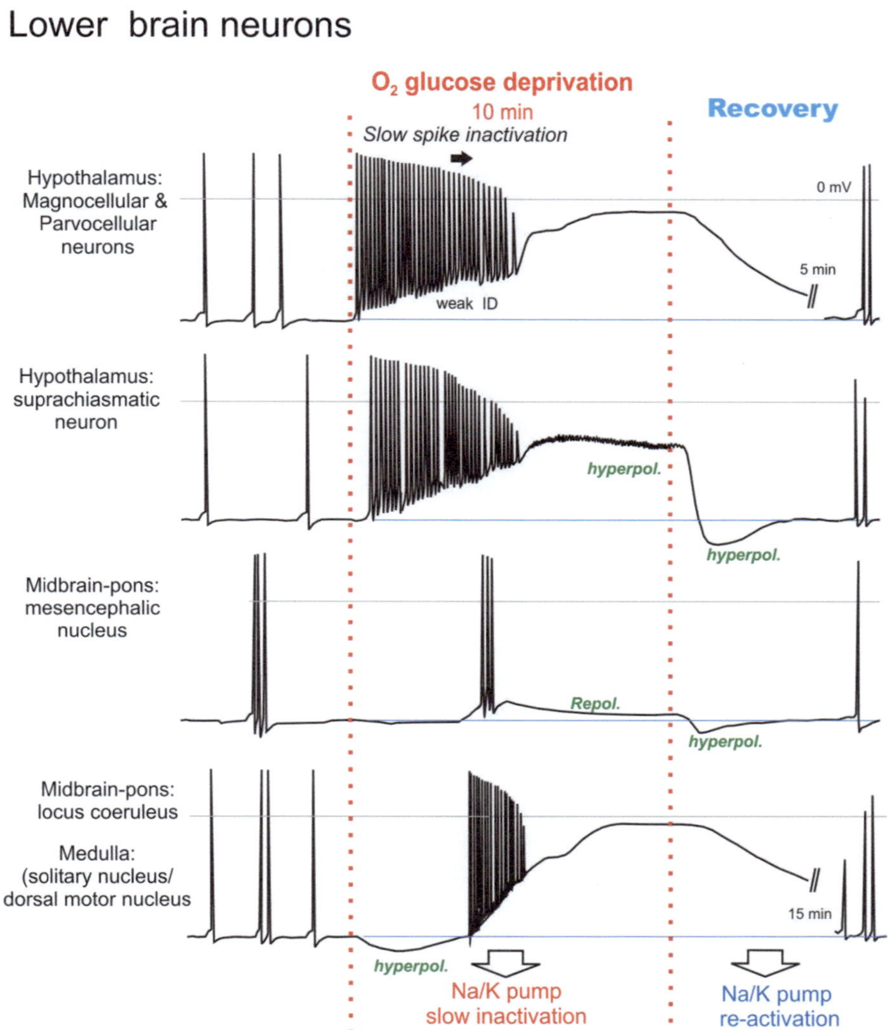

Fig. 10.7 Summary of lower neuron responses to simulated ischemia. In contrast to Fig. 10.6, all four general brainstem regions tested display neurons that undergo only gradual spike inactivation and comparatively gradual AD, here labelled ID (ischemic depolarization). Most importantly, these neurons substantially repolarize post-OGD. Of particular note, MES neurons maintain resting potential in the face of OGD. We propose that the these properties can be at least partially explained by brainstem and hypothalamic neurons possessing more resilient isoforms of the Na^+/K^+ pump than higher neurons

10.7 Discussion

Whether evoked by traumatic brain injury, near-drowning, cardiac arrest or pulmonary failure global ischemia deprives the *entire* brain of oxygen and glucose. In 1 year 400,000 North Americans experience sudden cardiac arrest. More than half of the survivors display permanent brain damage to varying degrees. Many are alive because somehow the brainstem resists the same ischemic period that injures the higher brain [5, 50]. However there is no evidence for a preferred or selective regional reperfusion of brainstem under any of these diverse situations. Also there are no data indicating that higher brain regions consume more energy per unit volume than lower regions. Nor is there any evidence that a cortical neuron requires more energy than a brainstem neuron nor that neurons from higher vs lower regions systematically differ in their ion channel composition. Yet the outcome from global ischemia is often diffuse damage to the higher brain with sparing of the hypothalamus and brainstem. As a result, over several weeks many afflicted patients transition from coma to a persistent vegetative state [3–5, 12–14]. Neocortex, striatum, hippocampus and thalamus are particularly vulnerable to ischemia [20, 35, 36, 38, 51] as is cerebellar cortex [28]. In the current study using brain slices, we document a terminal AD in these higher regions…but a weak and recoverable AD in all four hypothalamic nuclei and all four brainstem nuclei recorded.

Several studies using LT imaging in live cortical slices exposed to OGD show that, as the AD front passes by a recording electrode, there is a sudden mass depolarization of neurons. The cells do not repolarize based on non-reversible light scattering and lost evoked excitability [22, 23, 27, 29, 42, 43]. We image milder events in brainstem in the same coronal slices that support strong AD in hippocampus or cerebellar cortex. Dorsal brainstem nuclei display mini-fronts of AD but recover to baseline with nominal light scattering, indicating minimal neuronal damage. AD propagation is significantly slower and less damaging than higher gray regions with the exception of the superficial superior colliculus. These optical data suggest that the brainstem is comparatively protected from OGD. We investigated further using whole-patch recording.

OGD eliminates the evoked field potential in neocortical and hippocampal slices and this requires that the AD propagates through the recorded region [23, 29]. LT imaging reveals a similar vulnerability of striatal neurons in slices [13, 24, 43, 46] as also recorded in vivo in the ischemic core [20, 35, 36]. Pyramidal neurons have always been the prototype for CNS neurons under ischemic stress. Clearly PyN sensitivity is not representative of lower brain neurons which we show are slow to spike-inactivate and only depolarize gradually (below). Importantly, strong AD elicits vasoconstriction in neocortex (inverse neurovascular coupling) which exacerbates ischemic injury [19–21]. Whether weaker AD evokes less vasoconstriction in the brainstem is an important and open question.

Neurons of the hypothalamus, midbrain-pons and medulla easily survived 10 min of OGD. For example, locus ceruleus neurons and the mesencephalic neurons recovered 80–100 % of their membrane potential, input resistance and action potential amplitude. Moreover post-OGD, newly acquired neurons could be readily obtained in lower brain nuclei. Said neurons displayed membrane potentials, whole cell input resistances and action potential amplitudes similar to neurons *not* previously exposed to OGD [45]. MES neurons could even recover from multiple OGD exposures.

In every neuronal type we tested, ouabain exposure induced an OGD-like depolarization. However the hyperpolarizations recorded just after OGD exposure or removal were *not* observed with ouabain. Such hyperpolarizations are commonly attributed to activation of an ATP-sensitive K^+ current (K^+_{ATP}). As the ATP/ADP ratio drops, K^+_{ATP} opens but this would occur with OGD but not ouabain exposure. Alternately, the hyperpolarizations could be generated by up-regulating the Na^+/K^+ pump by OGD, a possible scenario given that the pump normally operates at only 50 % capacity [52]. While the K_{ATP} channel blocker glibenclamide has been considered specific, there is evidence that it can directly alter the Na^+/K^+ pump, thereby evoking hyperpolarization via pump activation [53] or evoking depolarization by pump inhibition [54]. Therefore early OGD-induced hyperpolarizations are not necessarily generated by K^+_{ATP} activation, but rather pump up-regulation.

Medullary neurons of dNTS and DMV also survive the OGD period that kills pyramidal and striatal neurons [39]. DNTS cells recovered >70 % of their mean membrane potential and action potential amplitude post-OGD. Newly patched neurons could be readily obtained in both DMV and dNTS that had recovered from OGD [45]. So in contrast to higher neurons which are quickly silenced by spike inactivation and depolarization, brainstem neurons depolarize slowly and are able to fire during both OGD onset and recovery. This gradual depolarization may activate brainstem neurons in cardio-respiratory centers that could aid a subject during hypoxia [55] or ischemia. Indeed, the LC projects to a number of brain regions involved in circulatory regulation including hypothalamus, nucleus of the solitary tract, nucleus ambiguous and dorsal motor nucleus of the vagus [55]. This neuronal resilience would help maintain vital functions during metabolic stress. The general response to OGD by each neuronal type in the current study is diagrammed in Figs. 10.6 and 10.7. Note the slow, delayed and recoverable AD generated by lower neurons, but not higher neurons.

10.7.1 A Crucial Role for the Types of Na⁺/K⁺ Pump **Isoform** *Expressed in Neurons?*

Although not emphasized in this chapter (but see Fig. 10.5a), exposure to 100 μM ouabain in each neuronal type tested evoked depolarization similar to that elicited by OGD. In higher neurons, ouabain induced rapid depolarization to near zero millivolts [27, 29, 46, 56] whereas hypothalamic and brainstem neurons only gradually depolarized and recovered. So as is consistently observed with intracellular recording [27] and imaging [22], ouabain exposure re-iterates the OGD response by inducing Na^+/K^+ pump failure. Recovery from 100 μM ouabain exposure is slower than OGD because ouabain only gradually unbinds the pump. Another consistent difference (noted above) is that any hyperpolarizations evoked during OGD are absent during ouabain exposure, possibly because ouabain inhibits up-regulation of the Na^+/K^+ pump.

So why are higher neurons more susceptible to ischemia than lower neurons? The presence of the α1 isoform of the Na^+/K^+ ATPase molecule increases vulnerability to ischemia compared to α2 and α3 versions [57, 58]. We have mined data (unpublished) from the Allen Brain Bank [59] indicating that these 'resilient' versions are proportionally most highly expressed in the hypothalamus and brainstem. It is interesting that motoneurons in spinal cord undergo strong AD [60], as well as expressing mRNA predominantly for the α1 isoform [61] as well as the protein itself [62].

The question arises as to why the rostral brain (and possibly spinal cord, Fig. 10.6) is programmed for ischemic vulnerability when ideally all neurons *should be* as resilient as possible. We propose that robust spreading depolarization is not simply a pathological mechanism, but is an active 'shutdown' process that has evolved in response to reduced blood flow caused by head trauma. This has conferred a survival advantage as vertebrates have evolved [63]. For example complete depolarization of a traumatized cortical area is preferable to a partial depolarization which can generate spastic or seizure activity. Upon head impact, immediately falling to the ground in an unconscious state reduces attention to the vulnerable individual, lessens the chance of further injury and lowers the metabolic rate, as well as reducing bleeding by lowering blood pressure and heart rate. A need to suppress hyperactivity would also hold for trauma to the spinal cord and so spinal motoneurons display strong AD during OGD [60]. Obviously a similar immediate and extensive shutdown of brainstem nuclei would prove fatal.

10.7.2 What Are the Molecular Events Linking Na^+/K^+ Pump Failure to Mass Neuronal Depolarization?

Despite much work by neurophysiologists, this question remains unanswered. The anoxic (or 'ischemic') depolarization is not simply the result of a slow run-down of the pump. Rather AD is driven by a formidable inward Na^+ current with K^+ efflux that accumulates extracellularly [24, 64, 65]. Yet blockers of standard Na^+ and K^+ channels only delay AD onset (ibid) and glutamate receptor activation is not required for AD except in cerebellar gray [17, 22, 24, 43, 65]. AD block requires a concentrated cocktail of ion-gated and ligand-gated channel antagonists that electrically silences the tissue [66–68]. The pannexin1 hemichannel [69] opens later during AD and appears not to be the primary conduit driving OGD-induced AD [66, 70] or spreading depression [71].

An unexplored possibility is that the Na^+/K^+ transporter *itself* is converted to an open channel that generates AD under ischemic conditions [72]. At nanomolar concentrations the marine poison *palytoxin* converts the ATPase molecule into an open cationic channel [73]. The resulting pore is wide enough to pass organic cations of <195 Da [74]. Held open, millions of Na^+ and K^+ ions/s are conducted per channel, whereas normally only ~100 ions/s are transported [75]. Given that there are thousands of Na^+/K^+ ATPase molecules in the plasma membrane of a single neuron, the resultant propagating current would be formidable, as evidenced by the AD-like response evoked by only 1 nM palytoxin [34, 72].

In summary, overriding the Na^+/K^+ pump using OGD, ouabain or palytoxin evokes steep and robust depolarization of neurons in higher, but not lower gray matter. Pump efficiency during ischemic stress may be an important factor in generating either strong or weak spreading depolarization by the brain. Understanding how brainstem neurons resist ischemia (perhaps by expressing pump isoforms that better function under OGD) can help us better understand why higher neurons are so exquisitely vulnerable to ischemia because clearly, lower neurons are not.

Acknowledgments Neuronal recordings were obtained by Dr. C. Devin Brisson as part of his doctoral thesis carried out in RDA's laboratory. Thanks to Dr. Susan Boehnke for statistical advice.

References

1. Multi-Society Task Force on PVS (1994) Medical aspects of the persistent vegetative state (1). N Engl J Med 330(21):1499–1508. doi:10.1056/NEJM199405263302107
2. Multi-Society Task Force on PVS (1994) Medical aspects of the persistent vegetative state (2). N Engl J Med 330(22):1572–1579. doi:10.1056/NEJM199406023302206
3. Kinney HC, Samuels MA (1994) Neuropathology of the persistent vegetative state. A review. J Neuropathol Exp Neurol 53(6):548–558
4. Adams JH, Graham DI, Jennett B (2000) The neuropathology of the vegetative state after an acute brain insult. Brain 123(Pt 7):1327–1338
5. Young GB (2009) Neurological prognosis after cardiac arrest. N Engl J Med 361:605–611
6. Falini A, Barkovich AJ, Calabrese G, Origgi D, Triulzi F, Scotti G (1998) Progressive brain failure after diffuse hypoxic ischemic brain injury: a serial MR and proton MR spectroscopic study. AJNR Am J Neuroradiol 19(4):648–652
7. Higuchi T, Graham SH, Fernandez EJ et al (1997) Effects of severe global ischemia on N-acetylaspartate and other metabolites in the rat brain. Magn Reson Med 37(6):851–857
8. Luigetti M, Goldsberry GT, Cianfoni A (2012) Brain MRI in global hypoxia-ischemia: a map of selective vulnerability. Acta Neurol Belg 112(1):105–107. doi:10.1007/s13760-012-0007-3
9. Laureys S, Owen AM, Schiff ND (2004) Brain function in coma, vegetative state, and related disorders. Lancet Neurol 3(9):537–546. doi:10.1016/S1474-4422(04)00852-X
10. Schiff ND, Ribary U, Moreno DR et al (2002) Residual cerebral activity and behavioural fragments can remain in the persistently vegetative brain. Brain 125(Pt 6):1210–1234
11. Bures J, Buresova O (1981) Cerebral [K+]e increase as an index of the differential susceptibility of brain structures to terminal anoxia and electroconvulsive shock. J Neurobiol 12(3):211–220. doi:10.1002/neu.480120303
12. Wytrzes LM, Chatrian GE, Shaw CM, Wirch AL (1989) Acute failure of forebrain with sparing of brain-stem function. electroencephalographic, multimodality evoked potential, and pathologic findings. Arch Neurol 46(1):93–97
13. Centonze D, Marfia GA, Pisani A et al (2001) Ionic mechanisms underlying differential vulnerability to ischemia in striatal neurons. Prog Neurobiol 63(6):687–696
14. Longstreth WT (2008) Neurological complications of cardiac arrest. In: Aminoff MJ (ed) Neurology and general medicine, 4th edn. Elsevier, Philadelphia, p 163
15. Jennett B (2005) Thirty years of the vegetative state: clinical, ethical and legal problems. Prog Brain Res 150:537–543. doi:10.1016/S0079-6123(05)50037-2

16. Beaumont JG, Kenealy PM (2005) Incidence and prevalence of the vegetative and minimally conscious states. Neuropsychol Rehabil 15(3–4):184–189. doi:10.1080/09602010443000489
17. Murphy TH, Li P, Betts K, Liu R (2008) Two-photon imaging of stroke onset in vivo reveals that NMDA-receptor independent ischemic depolarization is the major cause of rapid reversible damage to dendrites and spines. J Neurosci 28(7):1756–1772. doi:10.1523/JNEUROSCI.5128-07.2008
18. Farkas E, Bari F, Obrenovitch TP (2010) Multi-modal imaging of anoxic depolarization and hemodynamic changes induced by cardiac arrest in the rat cerebral cortex. Neuroimage 51(2):734–742. doi:10.1016/j. neuroimage. 2010.02.055
19. Dreier JP (2011) The role of spreading depression, spreading depolarization and spreading ischemia in neurological disease. Nat Med 17(4):439–447. doi:10.1038/nm.2333
20. Dijkhuizen RM, Beekwilder JP, van der Worp HB, Berkelbach van der Sprenkel JW, Tulleken KA, Nicolay K (1999) Correlation between tissue depolarizations and damage in focal ischemic rat brain. Brain Res 840(1–2):194–205
21. Kaminogo M, Suyama K, Ichikura A, Onizuka M, Shibata S (1998) Anoxic depolarization determines ischemic brain injury. Neurol Res 20(4):343–348
22. Jarvis CR, Anderson TR, Andrew RD (2001) Anoxic depolarization mediates acute damage independent of glutamate in neocortical brain slices. Cereb Cortex 11(3):249–259
23. Anderson TR, Jarvis CR, Biedermann AJ, Molnar C, Andrew RD (2005) Blocking the anoxic depolarization protects without functional compromise following simulated stroke in cortical brain slices. J Neurophysiol 93(2):963–979. doi:10.1152/jn.00654.2004
24. Tanaka E, Yamamoto S, Kudo Y, Mihara S, Higashi H (1997) Mechanisms underlying the rapid depolarization produced by deprivation of oxygen and glucose in rat hippocampal CA1 neurons in vitro. J Neurophysiol 78(2):891–902
25. Tanaka E, Yamamoto S, Inokuchi H, Isagai T, Higashi H (1999) Membrane dysfunction induced by in vitro ischemia in rat hippocampal CA1 pyramidal neurons. J Neurophysiol 81(4):1872–1880
26. Senatorov VV, Hu B (1997) Differential na(+)-K(+)-ATPase activity in rat lemniscal and non-lemniscal auditory thalami. J Physiol 502(Pt 2):387–395
27. Brisson CD, Lukewich MK, Andrew RD (2013) A distinct boundary between the higher brain's susceptibility to ischemia and the lower brain's resistance. PLoS One 8(11), e79589
28. Hamann M, Rossi DJ, Mohr C, Andrade AL, Attwell D (2005) The electrical response of cerebellar purkinje neurons to simulated ischaemia. Brain 128(Pt 10):2408–2420. doi:10.1093/brain/awh619
29. Brisson CD, Andrew RD (2012) A neuronal population in hypothalamus that dramatically resists acute ischemic injury compared to neocortex. J Neurophysiol 108(2):419–430. doi:10.1152/jn.00090.2012
30. Funke F, Kron M, Dutschmann M, Muller M (2009) Infant brain stem is prone to the generation of spreading depression during severe hypoxia. J Neurophysiol 10(5):2395–2410. doi:10.1152/jn.91260.2008
31. Karunasinghe RN, Lipski J (2013) Oxygen and glucose deprivation (OGD)-induced spreading depression in the substantia nigra. Brain Res 1527:209–221. doi:10.1016/j.brainres.2013.06.016
32. Richter F, Bauer R, Ebersberger A, Lehmenkuhler A, Schaible HG (2012) Enhanced neuronal excitability in adult rat brainstem causes widespread repetitive brainstem depolarizations with cardiovascular consequences. J Cereb Blood Flow Metab 32(8):1535–1545. doi:10.1038/jcbfm.2012.40
33. Brisson CD, Lukewich MK, Andrew RD (2013) A distinct boundary between the higher brain's susceptibility to ischemia and the lower brain's resistance. PLoS One 8(11):e79589. doi:10.1371/journal.pone.0079589
34. Brisson CD, Hsieh YT, Kim D, Jin AY, Andrew RD (2014) Brainstem neurons survive the identical ischemic stress that kills higher neurons: Insight to the persistent vegetative state. PLoS One 9(5):e96585. doi:10.1371/journal.pone.0096585
35. Memezawa H, Smith ML, Siesjo BK (1992) Penumbral tissues salvaged by reperfusion following middle cerebral artery occlusion in rats. Stroke 23(4):552–559

36. Koroleva VI, Vinogradova LV (2000) Ischemic and hypoxic depolarization in the rat neocortex. Zh Vyssh Nerv Deiat Im I P Pavlova 50(4):612–623
37. Andrew RD, Labron MW, Boehnke SE, Carnduff L, Kirov SA (2007) Physiological evidence that pyramidal neurons lack functional water channels. Cereb Cortex 17(4):787–802. doi:10.1093/cercor/bhk032
38. Sieber FE, Palmon SC, Traystman RJ, Martin LJ (1995) Global incomplete cerebral ischemia produces predominantly cortical neuronal injury. Stroke 26(11):2091–2095, discussion 2096
39. Martin RL (1999) Block of rapid depolarization induced by in vitro energy depletion of rat dorsal vagal motoneurones. J Physiol 519(Pt 1):131–141
40. Risher WC, Andrew RD, Kirov SA (2009) Real-time passive volume responses of astrocytes to acute osmotic and ischemic stress in cortical slices and in vivo revealed by two-photon microscopy. Glia 57(2):207–221. doi:10.1002/glia.20747
41. Feng G, Mellor RH, Bernstein M et al (2000) Imaging neuronal subsets in transgenic mice expressing multiple spectral variants of GFP. Neuron 28(1):41–51
42. Basarsky TA, Duffy SN, Andrew RD, MacVicar BA (1998) Imaging spreading depression and associated intracellular calcium waves in brain slices. J Neurosci 18(18):7189–7199
43. Joshi I, Andrew RD (2001) Imaging anoxic depolarization during ischemia-like conditions in the mouse hemi-brain slice. J Neurophysiol 85(1):414–424
44. Obeidat AS, Jarvis CR, Andrew RD (2000) Glutamate does not mediate acute neuronal damage after spreading depression induced by O2/glucose deprivation in the hippocampal slice. J Cereb Blood Flow Metab 20(2):412–422. doi:10.1097/00004647-200002000-00024
45. Andrew RD, Hsieh Y-T, Brisson CD (2012) Projection neurons in brainstem and hypothalamus intrinsically resist acute stroke injury while projection neurons in cortex, striatum and thalamus die. *Soc Neurosci Abstracts 2012*
46. White SH, Brisson CD, Andrew RD (2012) Examining protection from anoxic depolarization by the drugs dibucaine and carbetapentane using whole cell recording from CA1 neurons. J Neurophysiol 107(8):2083–2095. doi:10.1152/jn.00701.2011
47. Moriguchi S, Watanabe S, Kita H, Nakanishi H (2002) Enhancement of N-methyl- D-aspartate receptor-mediated excitatory postsynaptic potentials in the neostriatum after methamphetamine sensitization. an in vitro slice study. Exp Brain Res 144(2):238–246. doi:10.1007/s00221-002-1039-3
48. Yoshida S, Oka H (1998) Membrane properties of dissociated trigeminal mesencephalic neurons of the adult rat. Neurosci Res 30(3):227–234
49. Cowan AI, Martin RL (1992) Ionic basis of membrane potential changes induced by anoxia in rat dorsal vagal motoneurones. J Physiol 455:89–109
50. Young GB, Owen AM, Estraneo A, Moretta P, Trojano L (2013) Predictors of recovery of responsiveness in prolonged anoxic vegetative state. Neurology 81(14):1274–1275. doi:10.1212/WNL.0b013e3182a7ae28
51. Kawai K, Nitecka L, Ruetzler CA et al (1992) Global cerebral ischemia associated with cardiac arrest in the rat: I. dynamics of early neuronal changes. J Cereb Blood Flow Metab 12(2):238–249. doi:10.1038/jcbfm.1992.34
52. Gottron MA, Lo DC (2009) The Na^{+}/K^{+}-ATPase as a drug target for ischemic stroke. In: Annunziato L (ed) New strategies in stroke intervention. Humana Press, p 129. doi:10.1007/978-1-60761-280-3_8
53. Liu L, Gable ME, Garlid KD, Askari A (2007) Interactions of K+ATP channel blockers with na+/K+−ATPase. Mol Cell Biochem 306(1–2):231–237. doi:10.1007/s11010-007-9574-7
54. Lee SY, Lee CO (2005) Inhibition of na+−K+ pump and L-type Ca2+ channel by glibenclamide in guinea pig ventricular myocytes. J Pharmacol Exp Ther 312(1):61–68. doi:10.1124/jpet.104.074369
55. Yang JJ, Chou YC, Lin MT, Chiu TH (1997) Hypoxia-induced differential electrophysiological changes in rat locus coeruleus neurons. Life Sci 61(18):1763–1773
56. Balestrino M, Young J, Aitken P (1999) Block of (na+, K+)ATPase with ouabain induces spreading depression-like depolarization in hippocampal slices. Brain Res 838(1–2):37–44
57. Dobretsov M, Stimers JR (2005) Neuronal function and alpha3 isoform of the na/K-ATPase. Front Biosci 10:2373–2396

58. Blanco G (2005) Na, K-ATPase subunit heterogeneity as a mechanism for tissue-specific ion regulation. Semin Nephrol 25(5):292–303. doi:10.1016/j.semnephrol.2005.03.004
59. Hawrylycz MJ, Lein ES, Guillozet-Bongaarts AL et al (2012) An anatomically comprehensive atlas of the adult human brain transcriptome. Nature 489(7416):391–399. doi:10.1038/nature11405
60. Nohda K, Nakatsuka T, Takeda D et al (2007) Selective vulnerability to ischemia in the rat spinal cord: a comparison between ventral and dorsal horn neurons. Spine (Phila Pa 1976) 32(10):1060–1066. doi:10.1097/01.brs.0000261560.53428.90
61. Edwards IJ, Bruce G, Lawrenson C et al (2013) Na+/K+ ATPase alpha1 and alpha3 isoforms are differentially expressed in alpha- and gamma-motoneurons. J Neurosci 33(24):9913–9919. doi:10.1523/JNEUROSCI.5584-12.2013
62. Richards KS, Bommert K, Szabo G, Miles R (2007) Differential expression of Na+/K+-ATPase alpha-subunits in mouse hippocampal interneurones and pyramidal cells. J Physiol 585(Pt 2):491–505. doi:10.1113/jphysiol.2007.144733
63. Andrew RD, Brisson CD (2011) Trauma trumps stroke: why is our higher brain inept at dealing with blocked blood flow? Vasospasm 2011: The 11th International Conference on Neurovascular Events after Subarachnoid Hemorrhage, Cincinatti, pp 311–312
64. Czeh G, Aitken PG, Somjen GG (1992) Whole-cell membrane current and membrane resistance during hypoxic spreading depression. Neuroreport 3(2):197–200
65. Muller M, Somjen GG (2000) Na(+) dependence and the role of glutamate receptors and na(+) channels in ion fluxes during hypoxia of rat hippocampal slices. J Neurophysiol 84(4):1869–1880
66. Madry C, Haglerod C, Attwell D (2010) The role of pannexin hemichannels in the anoxic depolarization of hippocampal pyramidal cells. Brain 133(Pt 12):3755–3763. doi:10.1093/brain/awq284
67. Rossi DJ, Oshima T, Attwell D (2000) Glutamate release in severe brain ischaemia is mainly by reversed uptake. Nature 403(6767):316–321. doi:10.1038/35002090
68. Muller M, Somjen GG (1998) Inhibition of major cationic inward currents prevents spreading depression-like hypoxic depolarization in rat hippocampal tissue slices. Brain Res 812(1–2):1–13
69. Thompson RJ, Zhou N, MacVicar BA (2006) Ischemia opens neuronal gap junction hemichannels. Science 312(5775):924–927. doi:10.1126/science.1126241
70. Bargiotas P, Krenz A, Hormuzdi SG et al (2011) Pannexins in ischemia-induced neurodegeneration. Proc Natl Acad Sci U S A 108(51):20772–20777. doi:10.1073/pnas.1018262108
71. Karatas H, Erdener SE, Gursoy-Ozdemir Y et al (2013) Spreading depression triggers headache by activating neuronal Panx1 channels. Science 339(6123):1092–1095. doi:10.1126/science.1231897
72. Andrew RD, Britton R, McQueen SA, Kim D, Jin AY (2013) Could the failing na+/K+ pump convert to a channel during stroke? Soc Neurosci Abstract 337 07/T10
73. Artigas P, Gadsby DC (2004) Large diameter of palytoxin-induced na/K pump channels and modulation of palytoxin interaction by na/K pump ligands. J Gen Physiol 123(4):357–376. doi:10.1085/jgp.200308964
74. Rossini GP, Bigiani A (2011) Palytoxin action on the na(+), K(+)-ATPase and the disruption of ion equilibria in biological systems. Toxicon 57(3):429–439. doi:10.1016/j. toxicon.2010.09.011
75. Gadsby DC, Takeuchi A, Artigas P, Reyes N (2009) Review. peering into an ATPase ion pump with single-channel recordings. Philos Trans R Soc Lond B Biol Sci 364(1514):229–238. doi:10.1098/rstb.2008.0243
76. Dzhala V, Khalilov I, Ben-Ari Y, Khazipov R (2001) Neuronal mechanisms of the anoxia-induced network oscillations in the rat hippocampus in vitro. J Physiol 536(Pt 2):521–531

Chapter 11
Responsiveness and the Autonomic Control–CNS Two-Way Interaction in Disorders of Consciousness

Francesco Riganello

Abstract Functional neuroimaging has documented stimulus- or condition-related regional brain activation in subjects in VS/UWS or minimally conscious state and suggested retained connectivity in segregated cortical networks. It also provided indications of direct/indirect functional links between the autonomic control and higher brain activities. An integrated model (the central autonomic network) has been contrived to describe the interaction among neuronal structures involved in cognitive, sensory, or affective processes and autonomic regulation. The measures of heart rate variability (HRV) are thought to indirectly reflect these activities and interaction; integrative models equate HRV to autonomic nervous system outputs, with HRV reflecting affective, physiological, "cognitive," and behavioral elements and homeostatic responses. HRV provides tools for the evaluation of responsiveness in DoC and has proved reliable in physiological research in the absence of conscious behavior.

Keywords CAN • ANS • Responsiveness • Disorders of Consciousness • HRV

11.1 Introduction

The autonomic nervous system (ANS) compensates for internal and external needs to mediate in the processes that collectively sustain constancy of internal environment and guarantee adaptation and homeostasis [1, 2]. ANS can also mediate in patterns of brain activation at varying levels of complexity in affective, physiological, "cognitive", and behavioral processes. Neuroimaging research has documented bidirectional functional interactions of the autonomic control and activity in brain structures [3–8]. Research and systematic clinical application have expanded the

F. Riganello
Research in Advanced Neurorehabilitation (RAN), Institute S. Anna, Crotone, Italy
e-mail: f.riganello@istitutosantana.it

M.M. Monti, W.G. Sannita (eds.), *Brain Function and Responsiveness in Disorders of Consciousness*, DOI 10.1007/978-3-319-21425-2_11

scenario of the disorders of consciousness in raising questions about the underlying brain functions that may sustain responsiveness and conscious processes.

Specific measures of sympathetic activity are obtained from the cardiac norepinephrine spillover and microneurographic techniques or direct recording from the skeletal muscle [9, 10]. However, these methods are invasive, while only indirect methods are available to investigate the parasympathetic system [11, 12]. Indirect measures of the autonomic functional state and sympathetic/parasympathetic balance can be obtained by HRV methods, which have proved applicable in the description of changes in the adaptation to the momentary functional requirements.

11.2 The Autonomic Nervous System (ANS)

The ANS monitors and regulates a complex, highly differentiated network of distributed organs and biological sensors. It governs arterial blood pressure and regional blood flow in response to the tissues' metabolic demands, thermoregulation, motility and secretion of the gastrointestinal and respiratory tracts, micturition, and sexual function [13–16]. The autonomic functional state and its changes are the overall result of input from baroreceptors, chemoreceptors, atrial receptors, ventricular receptors, respiratory changes, vasomotor system changes, renin–angiotensin–aldosterone system balance, and thermoregulation [17, 18], with the main control by the sympathetic and parasympathetic branches [19].

11.2.1 Autonomic Nervous System and Heart Control

ANS mediates in the bidirectional communication between heart function and the Central Nervous System (CNS) [20]. Parasympathetic activity is predominant at rest with slower pace than the sinoatrial node intrinsic rate. This regulation depends on medullary centers, in particular the nucleus of solitary tract and rostroventrolateral medulla [21] that integrate sensory information from proprioceptors, chemoreceptors, and mechanoreceptors and from the telencephalic and limbic systems. The cardiovascular center adjusts the heart rate by operating on the relative balance between sympathetic and parasympathetic outflows [22]. The heart rate thus reflects the sympathetic/parasympathetic interplay with a dominant tonus at rest of parasympathetic n. vagus innervating the intrinsic cardiac nervous system and projecting to the sinoatrial node, atrioventricular node, and atrial cardiac muscle. Increased nerve activity triggers acetylcholine release and binding to muscarinic (mainly M_2) receptors, with decreasing rate of spontaneous depolarization in the sinoatrial and atrioventricular nodes and slowing heart rate. The effect of vagal stimulation is fast, with the heart rate quickly returning to its previous levels. Increased rate results from reduced vagal activity [23] or sympathetic activity above the intrinsic levels operated by the sinoatrial node [23].

11.2.2 Intrinsic Cardiac Network System

In mammals, the network of intracardiac neurons [24] is functionally flexible enough to qualify as an intrinsic quasi-nervous system [25]. The majority of vagal fibers is afferent; signals are sent to the brain to a greater extent than to any other major organ [26, 27]. Efferent sympathetic and parasympathetic activities are integrated with those of the heart intrinsic nervous system, including the afferent signals originating from mechanosensory and chemosensory neurons. Mechanical and hormonal information is processed into neurophysiological signals by sensory neurons [28, 29] and is transferred to the intrinsic nervous system and then to the brain [30, 31]. Information from mechanosensitive neurons travels through the glossopharyngeal and vagal nerves to the solitary tract nucleus, which connects with the other regulatory centers in the medulla to modulate the sympathetic outflow to heart and blood vessels. The parasympathetic outflow is modulated through connections to the dorsal vagal complex. Inputs from the heart also mediate in the release of hormones (oxytocin) and neurotransmitters (dopamine, adrenaline, noradrenaline) and in the neurohormonal interaction [32, 33]. Afferent information processed by the intrinsic cardiac nervous system [34] can influence activity in the frontocortical areas and motor cortex [5, 35], with modulation of attention, motivation [19, 36–38], perceptual sensitivity [39, 40], and emotional processing [5, 41, 42].

11.3 The CAN Model and the Autonomic/CNS Interaction

An integrated model has been proposed (usually referred to as the central autonomic network; CAN) [43] in which neuronal structures involved in cognitive, affective, and autonomic regulation are functionally linked to heart function. The interplay between autonomic control and the CNS is modeled as a functional setup connecting through feedback and feedforward loops the brainstem solitary tract nucleus with forebrain structures such as the anterior cingulate, insula, ventromedial prefrontal cortex, amygdala, and hypothalamus with bidirectional interactions between rostral and caudal systems. The encephalic structures identified in animal research and human fMRI studies are the parabrachial nucleus/locus ceruleus, cerebellum, periaqueductal gray, hypothalamus, amygdala, and insular and dorsomedial prefrontal cortices [44]. Some brain regions (dorsolateral prefrontal cortex, mediodorsal thalamus, hippocampus, caudate, septal nucleus, and middle temporal gyrus) have been suggested as possibly unique to humans [7]. All structures receive converging visceral and nociceptive inputs (as well as those from thermoreceptors and muscle receptors) and generate stimulus-specific patterns of autonomic responses via projections to preganglionic sympathetic and parasympathetic neurons [45, 46]. The medial prefrontal cortex projecting to interconnected limbic cortical regions and subcortical nuclei (insula, medial temporal lobes, amygdala, ventral striatum, ventral caudate and putamen, mediodorsal thalamus, and hypothalamus) is regarded as a key structure in the brain control of cardiovascular responses [47–49]. In the CAN model, the

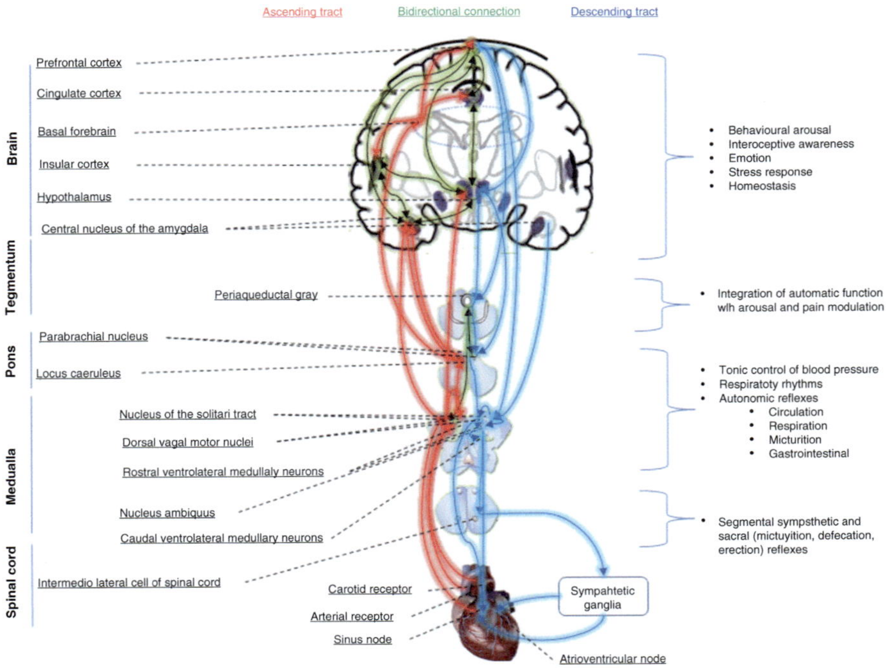

Fig. 11.1 Schematic outline of the central autonomic network model

ANS is thought to mediate homeostatic regulation through the interplay of the opposing but functionally reciprocal sympathetic and parasympathetic branches [43, 50, 51] and to be involved in the modulation of autonomic output in response to pain and emotional or behavioral stimulus conditions [8, 51, 52]. Structures involved in cognitive, affective, and autonomic regulation are functionally linked to heart function (Fig. 11.1). Neuroimaging has also outlined functional links between autonomic control and activity in brain structures involved in attention and conscious processes [5, 7, 51], consistent with the stimulus-related changes in the autonomic functional state observed in subjects otherwise diagnosed as being in Vegetative State/Unresponsive Wakefulness Syndrome (VS/UWS) [37, 53–56].

11.4 HRV Measures of Autonomic Function

The heart rate variations around the mean value over the time sample epoch mirror to a substantial extent of the cardiorespiratory control system and is now regarded as a reliable descriptor of the sympathetic/parasympathetic functional interplay and influence on heart function [57–59]. HRV measures are thought to reflect both ANS dynamics and heart–brain interaction and – to some extent and possibly only indirectly – also higher brain functions [3, 6, 60].

HRV recording techniques are noninvasive; HRV signals have an excellent signal-to-noise ratio compared to those investigated in neuroscience or clinical neurophysiology [61, 62]. However, the HRV signal is not periodic; its stimulus- or condition-related changes are within the physiological range in the absence of cardiac disorders and are seldom detectable without appropriate data treatment. To this purpose, the tachogram needs processing in the time or frequency domains or by geometric or nonlinear methods [59]. The HRV descriptors derived in the time domain quantify the variance during the recording window. The fast Fourier transform (FFT) or autoregressive models are of common use for analyses in the frequency domain. The signal spectral profile is characterized by three main components: (1) the high-frequency interval (0.15–0.5 Hz; HF) is mainly associated with activation of the parasympathetic nervous system; (2) the low-frequency interval (0.04–0.15 Hz; LF) is thought to reflect contributions from both the parasympathetic and sympathetic systems; and (3) the very-low-frequency bandwidth (<0.04 Hz; VLF) is thought to reflect temperature, vasomotor, hormonal, and metabolic regulation. Methods to describe the complexity, irregularity, or randomness of HRV and its changes (e.g., the approximate entropy (ApEn) or sample entropy (SampEn)) are available and in use [63, 64]. Stimulus- or condition-related changes occurring in selected frequency intervals of the HRV frequency spectrum are deemed expressive of specific functional changes [52].

11.5 Autonomic Control and Responsiveness in the Disorders of Consciousness

HRV has been successfully applied to study posttraumatic patients and subjects in VS/UWS or minimally conscious state. Improved HRV within 3 months from TBI was correlated with recovery of autonomic functions in a prospective study [65]. Sympathetic activation in response to auditory stimuli was reduced in acute VS/UWS even in subjects with a relatively good recovery [66]. Increased sympathetic response and recovery of the sympathovagal balance were correlated with recovery of consciousness [67]. In contrast, a severe sympathetic and motor overactivity syndrome indicative of over-responsiveness to afferent stimuli was associated with poor outcome [68].

VS/UWS subjects can process complex stimulus conditions, with greater responsiveness to emotional than nonemotional stimulus conditions [53]. Music appears peculiarly efficient in promoting arousal and responsiveness in DoC [69], with activation of the superior temporal gyri by music possibly predicting the evolution from VS/UWS [70]. Autonomic activation and the EEG amplitude reportedly increase in response to preferred music [69]. Changes in the HRV patterns of response indicative of enhanced activity of the cardiovascular system were observed after 14-day music therapy [71]. Replicable changes in the sympathovagal balance have been identified in VS/UWS during passive listening to symphonic music of different authorships by means of advanced data mining analyses of HRV descriptors. These descriptors allowed cluster of the autonomic responses as indicative of positive or negative emotions in both VS/UWS and awake posttraumatic subjects; the patterns of change were comparable across clinical conditions and among subjects for each music sample [56]. Electrodermal and HRV changes induced by the presence or voice of relatives (the *mom effect*) have been described in VS/UWS [72, 73]. The observation of autonomic changes in response to complex stimulus conditions with emotional value adds to the evidence from neuroimaging studies and has comparable implications as to the strategies and criteria of evaluation of the residual responsiveness of these subjects, their diagnosis and prognosis, and the planning of neurorehabilitation strategies [54].

11.6 Responsiveness in the Disorders of Consciousness and Variability

Diagnosis, monitoring of evolution, and early prognosis in DoC still rest mostly on the clinical evaluation of responsiveness [74]. However, the probability of observing an established clinical indicator can vary also in single subjects [75, 76]. For instance, the incidence of a visual pursuit response (a marker of evolution into the minimally conscious state also observed in about 20 % of VS/UWS subjects) was higher in the morning than in the afternoon testing during a 13-week neurorehabilitation protocol in both VS/UWS and minimally conscious state [77]. When subjects were repeatedly tested during the day, a response was most frequently detected at about 10.30 am and 3.00 pm and never at postprandial time [76]. The overall probability of observing a visual pursuit response at least once over the course of a single day was 30–40 % [76], in agreement with the reported rate of misdiagnosis between VS/UWS and the minimally conscious state [78, 79]. This variability of response could be accounted for by residual fragments of circadian/ultradian cycles asynchronous among subjects or by the fluctuations of the brain functional state that are observed in healthy humans and animals. Although apparently spontaneous, these fluctuations correlate with neuronal and non-neuronal biological parameters that vary within the physiological range depending on the momentary functional or homeostatic requirements and interact with each other [80]. Among these parameters are the spontaneous or environment-related changes in temperature, metabolism (e.g., glucose, ammonia), blood flow, oxygen extraction, pO_2, pCO_2, blood availability of ferritin, hormones (thyroxine, sexual hormones, cortisol, steroids, ATCH) and neurohormonal interaction, light–dark regulation of serotonin, histamine and dopamine/melatonin/vitamin B_{12} secretion and interaction, etc. These parameters can individually or collectively account for unexplained or underestimated individual variability and may become clinically relevant on occasions. The incidence of a pursuit response in VS/UWS subjects was found to correlate with and to be predicted by the HRV measures of sympathovagal balance and autonomic functional state. The HRV descriptor values at which the observation of a response was most frequent or unusual have been identified and the predictability of the model has been validated [81].

11.7 Comment

Extensive monitoring of the autonomic nervous system is advisable in the disorders of consciousness, and HRV techniques appear suitable of application in this field. HRV methodologies benefit from being noninvasive and with positive benefit/cost ratios, and measures are obtained at limited costs, labor, and accuracy of recording. Albeit indirect, information on the ANS functional state or responsiveness can be obtained also in absence of the subject's collaboration and when sophisticated experimental designs and data recording procedures are impracticable or difficult as is the case with the severe disorders of consciousness. Specifically, monitoring by HRV techniques is practicable with the same methodological approach in early phases at the ICU or hospital, during the neurorehabilitation procedures, and with remote control at home. The CAN model provides neuroscientists and clinicians with an independent approach to be used both in the understanding of the HRV descriptors of the autonomic–CNS interaction and supplementing clinical or neuroimaging observations.

Functional neuroimaging has identified brain activation suggesting awareness in only a small percentage of subjects [82]. The residual functions were otherwise regarded as resulting from intact but functionally disconnected cortical modules [74, 82], while the preservation of specific neural structures and residual resources has been suggested to vary with the heterogeneity of etiology and pathophysiology or extension and severity of brain damage [83]. The current clinical criteria for diagnosis and prognosis of the disorders of consciousness based on neurological signs of responsiveness should be reconsidered. The criteria should include the variability of responsiveness over time, and multiple testing should be mandatory to reduce the risks of misclassification. The correlation with the functional state of the ANS should be regarded as an independent indicator for diagnosis and prognosis and taken into proper account; monitoring should be extensive and focus also on non-neuronal factors. Reclassification of some patients or classes of patients may prove appropriate if based on systematic investigation.

References

1. Faith M, Thayer JF (2001) A dynamical systems interpretation of a dimensional model of emotion. Scand J Psychol 42:121–133
2. Posner J, Russell JA, Peterson BS (2005) The circumplex model of affect: an integrative approach to affective neuroscience, cognitive development, and psychopathology. Dev Psychopathol 17:715–734
3. Appelhans BM, Luecken LJ (2006) Heart rate variability as an index of regulated emotional responding. Rev Gen Psychol 10:229–240
4. Critchley HD, Mathias CJ, Josephs O, O'Doherty J, Zanini S, Dewar B-K, Cipolotti L, Shallice T, Dolan RJ (2003) Human cingulate cortex and autonomic control: converging neuroimaging and clinical evidence. Brain 126:2139–2152
5. Lane RD, McRae K, Reiman EM, Chen K, Ahern GL, Thayer JF (2009) Neural correlates of heart rate variability during emotion. Neuroimage 44:213–222

6. Mashin VA, Mashina MN (2000) Analysis of the heart rate variability in negative functional states in the course of psychological relaxation sessions. Hum Physiol 26:420–425
7. Napadow V, Dhond R, Conti G, Makris N, Brown EN, Barbieri R (2008) Brain correlates of autonomic modulation: combining heart rate variability with fMRI. Neuroimage 42: 169–177
8. Riganello F, Dolce G, Sannita W (2012) Heart rate variability and the central autonomic network in the severe disorder of consciousness. J Rehabil Med 44:495–501
9. Esler M (1993) Clinical application of noradrenaline spillover methodology: delineation of regional human sympathetic nervous responses. Pharmacol Toxicol 73:243–253
10. Wallin BG, Charkoudian N (2007) Sympathetic neural control of integrated cardiovascular function: insights from measurement of human sympathetic nerve activity. Muscle Nerve 36:595–614
11. Grossman P, Taylor EW (2007) Toward understanding respiratory sinus arrhythmia: relations to cardiac vagal tone, evolution and biobehavioral functions. Biol Psychol 74:263–285
12. Lehrer PM, Vaschillo E, Vaschillo B, Lu S-E, Eckberg DL, Edelberg R, Shih WJ, Lin Y, Kuusela TA, Tahvanainen KUO, Hamer RM (2003) Heart rate variability biofeedback increases baroreflex gain and peak expiratory flow. Psychosom Med 65:796–805
13. Fujibayashi M, Hamada T, Matsumoto T, Kiyohara N, Tanaka S, Kotani K, Egawa K, Kitagawa Y, Kiso Y, Sakane N, Moritani T (2009) Thermoregulatory sympathetic nervous system activity and diet-induced waist-circumference reduction in obese Japanese women. Am J Hum Biol 21:828–835
14. Kuusela TA, Kaila TJ, Kähönen M (2003) Fine structure of the low-frequency spectra of heart rate and blood pressure. BMC Physiol 3:11
15. Medford N, Critchley HD (2010) Conjoint activity of anterior insular and anterior cingulate cortex: awareness and response. Brain Struct Funct 214:535–549
16. van Orshoven NP, Oey PL, van Schelven LJ, Roelofs JMM, Jansen PA, Akkermans LMA (2004) Effect of gastric distension on cardiovascular parameters: gastrovascular reflex is attenuated in the elderly. J Physiol 555:573–583
17. Aubert AE, Seps B, Beckers F (2003) Heart rate variability in athletes. Sports Med Auckl NZ 33:889–919
18. Cooke WH, Cox JF, Diedrich AM, Taylor JA, Beightol LA, Ames JE, Hoag JB, Seidel H, Eckberg DL (1998) Controlled breathing protocols probe human autonomic cardiovascular rhythms. Am J Physiol 274:H709–H718
19. Riganello F, Garbarino S, Sannita WG (2012) Heart rate variability, homeostasis, and brain function: a tutorial and review of application. J Psychophysiol 26:178–203. doi:10.1027/0269-8803/a000080
20. Kawashima T (2005) The autonomic nervous system of the human heart with special reference to its origin, course, and peripheral distribution. Anat Embryol (Berl) 209:425–438
21. Spyer KM, Gourine AV (2009) Chemosensory pathways in the brainstem controlling cardiorespiratory activity. Philos Trans R Soc B Biol Sci 364:2603–2610
22. Shaffer F, Venner J (2013) Heart rate variability anatomy and physiology. Biofeedback 41: 13–25
23. Hainsworth R (1995) The control and physiological importance of heart rate. In: Malik M, Camm AJ (eds) Heart rate variability. Futura Publishing Company, Inc, Armonk
24. Verkerk AO, Remme CA, Schumacher CA, Scicluna BP, Wolswinkel R, de Jonge B, Bezzina CR, Veldkamp MW (2012) Functional Nav1.8 channels in intracardiac neurons: the link between SCN10A and cardiac electrophysiology. Circ Res 111:333–343
25. Armour JA (2008) Potential clinical relevance of the "little brain" on the mammalian heart. Exp Physiol 93:165–176
26. Berntson GG, Sarter M, Cacioppo JT (2003) Ascending visceral regulation of cortical affective information processing. Eur J Neurosci 18:2103–2109
27. Cameron OG (2009) Visceral brain–body information transfer. Neuroimage 47:787–794
28. Armour JA, Kember GC (2004) Cardiac sensory neurons. Basic and clinical neurocardiology, In: Armour JA, Ardell JL, (eds). Oxford University Press, New York. pp. 79–117

29. Cheng Z, Powley TL, Schwaber JS, Doyle FJ (1997) Vagal afferent innervation of the atria of the rat heart reconstructed with confocal microscopy. J Comp Neurol 381:1–17
30. Kukanova B, Mravec B (2006) Complex intracardiac nervous system. Bratisl Lek Listy 107:45
31. Randall DC, Brown DR, McGuirt AS, Thompson GW, Armour JA, Ardell JL (2003) Interactions within the intrinsic cardiac nervous system contribute to chronotropic regulation. Am J Physiol Regul Integr Comp Physiol 285:R1066–R1075
32. Clerico A, Giannoni A, Vittorini S, Passino C (2011) Thirty years of the heart as an endocrine organ: physiological role and clinical utility of cardiac natriuretic hormones. Am J Physiol Heart Circ Physiol 301:H12–H20
33. Randall DC, Evans JM, Billman GE, Ordway GA, Knapp CF (1981) Neural, hormonal and intrinsic mechanisms of cardiac control during acute coronary occlusion in the intact dog. J Auton Nerv Syst 3:87–99
34. Armour JA (1991) Intrinsic cardiac neurons. J Cardiovasc Electrophysiol 2:331–341
35. McCraty R, Shaffer F (2015) Heart rate variability: new perspectives on physiological mechanisms, assessment of self-regulatory capacity, and health risk. Glob Adv Health Med Improv Healthc Outcomes Worldw 4:46–61
36. Kaufmann T, Vögele C, Sütterlin S, Lukito S, Kübler A (2012) Effects of resting heart rate variability on performance in the P300 brain-computer interface. Int J Psychophysiol 83:336–341
37. Riganello F, Cortese MD, Dolce G, Sannita WG (2013) Visual pursuit response in the severe disorder of consciousness: modulation by the central autonomic system and a predictive model. BMC Neurol 13:164
38. Schandry R, Montoya P (1996) Event-related brain potentials and the processing of cardiac activity. Biol Psychol 42:75–85
39. MacKinnon S, Gevirtz R, McCraty R, Brown M (2013) Utilizing heartbeat evoked potentials to identify cardiac regulation of vagal afferents during emotion and resonant breathing. Appl Psychophysiol Biofeedback 38:241–255
40. Park H-D, Correia S, Ducorps A, Tallon-Baudry C (2014) Spontaneous fluctuations in neural responses to heartbeats predict visual detection. Nat Neurosci 17:612–618
41. Palma J-A, Benarroch EE (2014) Neural control of the heart: recent concepts and clinical correlations. Neurology 83:261–271
42. Taggart P, Boyett MR, Logantha SJRJ, Lambiase PD (2011) Anger, emotion, and arrhythmias: from brain to heart. Front Physiol 2:67
43. Benarroch EE (1993) The central autonomic network: functional organization, dysfunction, and perspective. Mayo Clin Proc 68:988–1001
44. Benarroch EE (2007) Enteric nervous system: functional organization and neurologic implications. Neurology 69:1953–1957
45. Benarroch EE (2007) The autonomic nervous system: basic anatomy and physiology. Contin Lifelong Learn Neurol 13:13–32
46. Saper CB (2002) The central autonomic nervous system: conscious visceral perception and autonomic pattern generation. Annu Rev Neurosci 25:433–469
47. Bandler R, Keay KA, Floyd N, Price J (2000) Central circuits mediating patterned autonomic activity during active vs. passive emotional coping. Brain Res Bull 53:95–104
48. Barbas H, Saha S, Rempel-Clower N, Ghashghaei T (2003) Serial pathways from primate prefrontal cortex to autonomic areas may influence emotional expression. BMC Neurosci 4:25
49. Hsu DT, Price JL (2007) Midline and intralaminar thalamic connections with the orbital and medial prefrontal networks in macaque monkeys. J Comp Neurol 504:89–111
50. Thayer JF, Lane RD (2000) A model of neurovisceral integration in emotion regulation and dysregulation. J Affect Disord 61:201–216
51. Thayer JF, Lane RD (2009) Claude Bernard and the heart–brain connection: further elaboration of a model of neurovisceral integration. Neurosci Biobehav Rev 33:81–88
52. Friedman BH (2007) An autonomic flexibility–neurovisceral integration model of anxiety and cardiac vagal tone. Biol Psychol 74:185–199

53. Gutiérrez J, Machado C, Estévez M, Olivares A, Hernández H, Perez J, Beltrán C, Leisman G (2010) Heart rate variability changes induced by auditory stimulation in persistent vegetative state. Int J Disabil Hum Dev 9:357–362
54. Riganello F, Candelieri A, Quintieri M, Conforti D, Dolce G (2010) Heart rate variability: an index of brain processing in vegetative state? An artificial intelligence, data mining study. Clin Neurophysiol 121:2024–2034
55. Riganello F, Candelieri A, Quintieri M, Dolce G (2010) Heart rate variability, emotions, and music. J Psychophysiol 24:112–119
56. Riganello F, Quintieri M, Candelieri A, Conforti D, Dolce G (2008) Heart rate response to music: an artificial intelligence study on healthy and traumatic brain-injured subjects. J Psychophysiol 22:166–174
57. Lanfranchi PA, Somers VK (2002) Arterial baroreflex function and cardiovascular variability: interactions and implications. Am J Physiol Regul Integr Comp Physiol 283:R815–R826
58. Lombardi F (2002) Clinical implications of present physiological understanding of HRV components. Card Electrophysiol Rev 6:245–249
59. Task Force of the European Society of Cardiology and the North American Society of Pacing and Electrophysiology (1996) Heart rate variability: standards of measurement, physiological interpretation and clinical use. Circulation 93:1043–1065
60. Frazier TW, Strauss ME, Steinhauer SR (2004) Respiratory sinus arrhythmia as an index of emotional response in young adults. Psychophysiology 41:75–83
61. Ge D, Srinivasan N, Krishnan SM (2002) Cardiac arrhythmia classification using autoregressive modeling. Biomed Eng Online 1:5
62. Kleiger RE, Stein PK, Bigger JT (2005) Heart rate variability: measurement and clinical utility. Ann Noninvasive Electrocardiol 10:88–101
63. Fojt O, Holcik J (1998) Applying nonlinear dynamics to ECG signal processing. IEEE Eng Med Biol Mag 17:96–101
64. Richman JS, Moorman JR (2000) Physiological time-series analysis using approximate entropy and sample entropy. Am J Physiol Heart Circ Physiol 278:H2039–H2049
65. Su C-F, Kuo TB, Kuo J-S, Lai H-Y, Chen HI (2005) Sympathetic and parasympathetic activities evaluated by heart-rate variability in head injury of various severities. Clin Neurophysiol Off J Int Fed Clin Neurophysiol 116:1273–1279
66. Turkstra LS (1995) Electrodermal response and outcome from severe brain injury. Brain Inj 9:61–80
67. Wijnen VJM, Heutink M, van Boxtel GJM, Eilander HJ, de Gelder B (2006) Autonomic reactivity to sensory stimulation is related to consciousness level after severe traumatic brain injury. Clin Neurophysiol 117:1794–1807
68. Baguley IJ, Nott MT, Slewa-Younan S, Heriseanu RE, Perkes IE (2009) Diagnosing dysautonomia after acute traumatic brain injury: evidence for overresponsiveness to afferent stimuli. Arch Phys Med Rehabil 90:580–586
69. O'Kelly J, James L, Palaniappan R, Taborin J, Fachner J, Magee WL (2013) Neurophysiological and behavioral responses to music therapy in vegetative and minimally conscious states. Front Hum Neurosci 7:884
70. Okumura Y, Asano Y, Takenaka S, Fukuyama S, Yonezawa S, Kasuya Y, Shinoda J (2014) Brain activation by music in patients in a vegetative or minimally conscious state following diffuse brain injury. Brain Inj 28:944–950
71. Lee Y-C, Lei C-Y, Shih Y-S, Zhang W-C, Wang H-M, Tseng C-L, Hou MC, Chiang H-Y, Huang S-C (2011) HRV response of vegetative state patient with music therapy. Conf Proc Annu Int Conf IEEE Eng Med Biol Soc IEEE Eng Med Biol Soc Conf 2011:1701–1704
72. Dolce G, Riganello F, Quintieri M, Candelieri A, Conforti D (2008) Personal interaction in the vegetative state: a data-mining study. J Psychophysiol 22:150–156
73. Machado C, Korein J, Aubert E, Bosch J, Alvarez MA, Rodríguez R, Valdés P, Portela L, García M, Pérez N, Chinchilla M, Machado Y, Machado Y (2007) Recognizing a mother's voice in the persistent vegetative state. Clin EEG Neurosci 38:124–126

74. Celesia GG, Sannita WG (2013) Can patients in vegetative state experience pain and have conscious awareness? Neurology 80:328–329
75. Bosco A, Lancioni GE, Belardinelli MO, Singh NN, O'Reilly MF, Sigafoos J (2010) Vegetative state: efforts to curb misdiagnosis. Cogn Process 11:87–90
76. Candelieri A, Cortese MD, Dolce G, Riganello F, Sannita WG (2011) Visual pursuit: within-day variability in the severe disorder of consciousness. J Neurotrauma 28:2013–2017
77. Cortese MD, Riganello F, Arcuri F, Pugliese ME, Lucca LF, Dolce G, Sannita WG (2015) Coma recovery scale-R: variability in the disorder of consciousness. BMC Neurol (in press)
78. Andrews K, Murphy L, Munday R, Littlewood C (1996) Misdiagnosis of the vegetative state: retrospective study in a rehabilitation unit. BMJ 313:13–16
79. Schnakers C, Vanhaudenhuyse A, Giacino J, Ventura M, Boly M, Majerus S, Moonen G, Laureys S (2009) Diagnostic accuracy of the vegetative and minimally conscious state: clinical consensus versus standardized neurobehavioral assessment. BMC Neurol 9:35
80. Sannita WG (2006) Individual variability, end-point effects and possible biases in electrophysiological research. Clin Neurophysiol 117:2569–2583
81. Riganello F, Cortese MD, Dolce G, Lucca LF, Sannita WG (2015) The autonomic system functional state predicts responsiveness in DOC. J Neurotrauma 32(14): 1071-1077
82. Monti MM, Laureys S, Owen AM (2010) The vegetative state. BMJ 341:c3765
83. Gibson RM, Fernández-Espejo D, Gonzalez-Lara LE, Kwan BY, Lee DH, Owen AM, Cruse D (2014) Multiple tasks and neuroimaging modalities increase the likelihood of detecting covert awareness in patients with disorders of consciousness. Front Hum Neurosci 8:950

Chapter 12
The Medical Practice Impact of Functional Neuroimaging Studies in Patients with Disorders of Consciousness

James L. Bernat

Abstract Functional neuroimaging studies of patients with disorders of consciousness have shown a small number of patients clinically diagnosed as in a vegetative state who, by fMRI or FDG-PET activation patterns, retain ideational capacity to follow commands showing they are aware. This finding impacts medical practice in several ways. The diagnostic criteria for disorders of consciousness must evolve from purely clinical to clinical plus functional neuroimaging. Prognosis is affected because the subgroup who retain this capacity comprise those with a higher likelihood for spontaneous clinical improvement. A few patients can be taught communication systems using fMRI responses, improving patient-centered care by allowing their participation in their own medical treatment decisions involving neurorehabilitation, stimulation, and palliation. A few medicolegal practice issues also are impacted by fMRI studies. But before functional neuroimaging studies can be fully incorporated into medical practice in these patients, they must be better understood, validated, and standardized.

Keywords Consciousness • Functional neuroimaging • fMRI • FDG-PET • Vegetative state • Minimally conscious state • Locked-in syndrome

12.1 Introduction

Studies of functional neuroimaging and brain electrical potential recordings on patients with disorders of consciousness (DoC) are revolutionizing our understanding of the pathophysiology of human conscious awareness [1–3]. These studies raise practical questions including how neurologists and neurosurgeons should inform patients' family members about the studies and how physicians caring for DoC patients should incorporate them into their practices. While most commentators

J.L. Bernat, MD
Department of Neurology, Geisel School of Medicine at Dartmouth, Hanover, NH, USA
e-mail: bernat@dartmouth.edu

M.M. Monti, W.G. Sannita (eds.), *Brain Function and Responsiveness in Disorders of Consciousness*, DOI 10.1007/978-3-319-21425-2_12

classify the individual case and small series reports of clinical-radiographic and clinical-electroencephalographic correlations as neuroscience research [4], after discovering them on an Internet search, family members of unconscious patients often ask to have the tests conducted. Although these tests are categorically limited by their availability and technical constraints, clinicians must consider their place in practice to optimally respond to these requests.

Scholars who have analyzed the ethical aspects of incorporating these studies into clinical practice have reached several conclusions. First, clinicians must strike an overall balance between protecting DoC patients against the misuse of scientific information and maintaining research to understand consciousness and treatment of DoC [5]. Clinicians should share relevant neuroscience data with patients' families, but how and to what extent remains unclear. Illes and colleagues [6] have generated a list of frequently asked questions about functional neuroimaging studies and their limitations that they distribute to patients' family members to educate them. Graham and colleagues [7] argued that clinicians should disclose relevant neuroscience research results with families when the studies fulfill four criteria: they are scientifically valid, informative, reasonably reliable, and offer considerable benefit for the patient.

In this chapter, I discuss how physicians should incorporate recent neuroimaging and EEG-evoked potential studies in DoC into their patients' care. I discuss several circumstances in which physicians might include the results of these studies: formulating the patient's diagnosis and prognosis; communicating with family members; fostering patient-centered decision-making with the patient's lawful surrogate decision-maker; offering medical treatments including stimulation, neurorehabilitation, and palliative care; and in a few medicolegal situations. Several of these practical issues are discussed in greater detail elsewhere [8, 9]. General reviews of DoC are available [2, 10–12].

12.2 Case Reports and Studies

The emerging studies uncovering occult conscious awareness in patients with DoC can be grouped into two general categories: functional neuroimaging (particularly fMRI and FDG-PET) and recording of brain electrical potentials. The functional neuroimaging studies have been more dramatic, provocative, and convincing, but the brain electrical potential recording studies have the advantage of being available more widely.

Owen and colleagues first showed that fMRI could reliably reveal evidence of conscious awareness in one patient in a vegetative state (VS) even when a careful neurological examination could not [13]. The publication of subsequent case reports and small series of VS patients confirmed and expanded his original findings [14–17]. The experimental paradigm had a unique feature compared to those used in earlier studies. Unlike previous passive fMRI paradigms, they instructed the patient to imagine performing a particular motor activity, such as playing tennis or walking through the house. Patients so instructed who retained the capacity to perform the ideational task showed focal regions of evoked increases in blood-oxygen-level-dependent (BOLD) fMRI signal representing increased blood flow in brain premotor areas. They termed the fMRI responses resulting from this ideational task "willful modulation of brain activity" [16]. The subset of patients who retain this capacity of "top-down cognitive functioning" have been shown to possess greater connectivity between the anterior thalamus and prefrontal cortex [18].

Although most investigators have accepted that the fMRI findings of willful modulation of brain activity indicated the presence of at least some degree of conscious awareness, some knowledgeable critics have expressed skepticism about the quality or extent of the patients' awareness that these findings represented [19, 20] or questioned whether the investigators had "conflated findings *associated* with willful consciousness with the actual *occurrence* of willful consciousness" [21]. Other skeptics, with experience performing fMRI studies in DoC, have emphasized the formidable technical hurdles that must be overcome for researchers to obtain consistent and reliable fMRI data on which to draw firm conclusions, because of inter-examination interval variations in patient arousal state, motion, medication effects, and other variables affecting signal acquisition and analysis [14, 22]. To establish consistent and reliable data, repeated fMRI examinations are just as necessary as are repeated clinical examinations to validly assess the presence or absence of conscious awareness [3].

Several investigators have made the analogous claim of the technical capacity to detect covert awareness in patients with DoC using sophisticated EEG [23, 24], preserved EEG organization [25], or evoked potential recordings [26, 27] as well as the capacity to use bispectral analysis recordings of brain electrical potentials on normal people during general anesthesia for the same purpose [28]. Some of these reports have been controversial. For example, following the publicized report by Cruse and colleagues [29] of a bedside EEG device that could detect awareness in some VS patients, Goldfine and colleagues [30] reanalyzed the investigators' data and concluded that the published findings were invalid, a claim that was rebutted by the investigators.

12.3 Diagnosis

The most direct effect of the emerging fMRI reports showing the presence of awareness in a few patients with the clinical diagnosis of VS is that it changes the diagnosis in those patients because evidence for awareness must be totally absent for a diagnosis of VS. The accepted diagnostic criteria for VS are clinical and are delineated mostly in the negative, i.e., stipulating those activities that VS patients cannot perform [31]. Delineating diagnostic criteria by negative capacities encourages false-positive diagnostic errors because reasons other than the brain damage may account for why patients cannot perform certain tasks. For example, unconscious patients may be tested improperly without using validated assessment scales, they may be sedated by medications treating seizures and other conditions, or they may have medical comorbidities such as infection or dehydration that produce a toxic encephalopathy suppressing responsiveness [32]. These factors and the nature of the negative delineation of the diagnostic criteria for VS help explain why false-positive diagnostic errors have been reported in 40 % of patients in VS [33].

If we accept that the fMRI data showing willful modulation of brain activity are valid in certain cases of VS, then the original diagnosis must be altered because evidence of awareness invalidates the diagnosis of VS. Prior to these reports, neurologists believed that a careful neurological examination using an optimal clinical assessment scale for detecting conscious awareness – such as the Coma Recovery Scale-Revised (CRS-R) [34] – constituted the gold standard for diagnosis. Now it appears that a small subset of patients diagnosed as VS according to accepted clinical criteria have been misdiagnosed and should be correctly categorized as a minimally conscious state (MCS) rather than VS [35].

The false-positive diagnoses of VS raise the larger question of whether the clinical diagnostic criteria for VS should be revised to incorporate functional neuroimaging data and for fMRI or FDG-PET to become a routine part of clinical assessment of these patients, as Coleman and colleagues [36] have advocated. I agree with them that over the next several years, the diagnostic criteria of VS should evolve from being purely clinical to clinical plus functional neuroimaging. But before this can occur, more systematic studies must be conducted to better delineate and standardize the fMRI and FDG-PET findings. Clinical syndromes such as VS and MCS have been carved out of the continuum of diffuse brain damage. The precise boundaries of these syndromes will likely need adjustment as the functional neuroimaging and other studies provide greater insights into the anatomy of consciousness and the pathophysiology of DoC.

In the largest study of functional neuroimaging in DoC, Stender et al. [37] showed that FDG-PET patterns had a higher sensitivity for detecting awareness among MCS patients (93 %) than fMRI (45 %) and also a higher congruence with the findings of awareness on clinical examination using the CRS-R (85 % vs. 63 %). They found that 32 % of their clinically unresponsive patients had either fMRI or FDG-PET evidence for at least minimal awareness. They recommended that a complete evaluation for occult awareness in DoC patients should include both of these functional neuroimaging modalities and the CRS-R [37].

12.4 Prognosis

Whether patients in VS or MCS will later improve or are destined to remain in that state chronically is an essential determination for clinicians. The Multi-Society Task Force on PVS [31] concluded that once a patient had been in VS from a nontraumatic etiology for more than 3 months or from traumatic brain injury (TBI) for more than a year, there was only a slight change for significant improvement subsequently. More recent data have shown these predictions to be overly pessimistic, particularly in the TBI subgroup. Several reports have described rare but more frequent late spontaneous improvements in the level of consciousness of some VS patients [38–41].

Beginning with the initial report of Owen et al. [13], several investigators have reported that the small subset of VS patients showing fMRI patterns of willful modulation of brain activity also were those who later were most likely to show spontaneous clinical improvement. Owen's initial patient, who was clinically in VS at 5 months after TBI and who showed fMRI evidence of awareness by willful modulation, improved spontaneously to show clinical signs of awareness at 11 months after injury. Other investigators have made similar observations about a few patients in the subset of VS patients who demonstrated willful modulation [18] or who showed characteristic fMRI responses to hearing the subject's own name spoken [42]. In the large recent study of functional neuroimaging in DoC patients, FDG-PET correctly predicted outcome in 74 %, while fMRI prediction was correct in 56 % [37].

It has been known for decades that the outcomes of patients in VS resulting from TBI were more favorable than patients in VS resulting from hypoxic-ischemic neuronal damage. Most of the VS patients who have been reported to demonstrate willful modulation and who later spontaneously improved were in VS as a result of TBI and not hypoxic-ischemic neuronal injury. Similarly, studies measuring EEG responses of MCS patients to ideational tasks show positive activation in most patients who had TBI but in none who had hypoxic-ischemic injuries [23]. Yet, some studies also have shown activation responses in hypoxic-ischemic cases. Monti and colleagues [16] found that six of ten DoC patients who were able to respond to an executive function task were from non-traumatic brain injury, including two of three VS patients who had suffered hypoxic-ischemic injury. Similarly, John and colleagues [43] reported one post-cardiac arrest hypoxic-ischemic VS patient in whom QEEG showed the ability to perform a mental imagery task. These small numbers show that although DoC patients with hypoxic-ischemic brain injuries may be among those detected to have occult awareness, they remain the minority of reported cases.

Many TBI patients develop VS as a result of diffuse axonal injury (DAI), a serious condition in which rotational traumatic forces shear axons diffusely at the junction between the cerebral gray matter and white matter. Although DAI often causes a severe brain injury with a poor outcome, because neurons usually remain intact, it is generally not as irreversible as a hypoxic-ischemic neuronal injury [44, 45]. The most dramatic example is the unprecedented recovery of a patient in MCS for 19 years before he began speaking and interacting in whom diffusion tensor imaging showed findings interpreted as axonal regrowth in the posterior parietal white matter [46]. Both FDG-PET and fMRI findings may have prognostic value among TBI patients (and in some hypoxic-ischemic patients) who demonstrate willful modulation because they may serve as a marker for subsequent spontaneous clinical improvement.

12.5 Communication

Once otherwise unresponsive patients with willful modulation of brain activity could be identified, investigators made attempts to establish systems to allow the patients to communicate by using their fMRI responses. One of the patients described by Monti and colleagues [16] was taught to reliably respond "yes" or "no" to binary questions by assigning the tennis-playing ideational paradigm to indicate "yes" and the room-walking ideational paradigm to indicate "no." The inability of an aware patient to communicate (particularly one wrongly believed to be unconscious) obviously produces profound suffering. Establishing a reliable method for the patient to communicate is essential to mitigate their suffering, to acknowledge that the medical and nursing staff knows they are aware, and to allow patients to express themselves, to optimize neurorehabilitation and palliative care, and possibly to participate in medical decision-making.

Investigators have tested MCS patients who remain aware but profoundly unresponsive to test various strategies of communication. Recognition of the patient's own name was found to be the most reliable test, yielding 100 % accuracy in the absence of a coexisting confusional state in a study of 144 patients in MCS after TBI [47]. Using fMRI responses, Naci and Owen [48], using a convenience sample of three DoC patients, found that one VS patient and one of two MCS patients could successfully use selective auditory attention to convey their ability to follow commands and to communicate.

Despite these encouraging findings, there remain serious barriers to using fMRI-evoked changes in response to ideational paradigms as responses for communication, particularly in obtaining consistent and reliable responses. Very few of the patients showing willful modulation have been able to achieve successful communication systems. Patients with VS and MCS, especially those from TBI, often also have aphasia, sedation from anticonvulsant drugs, global cognitive impairment from diffuse brain damage, and confusion from medical comorbidities. Each of these factors diminishes reliable responsiveness that makes communication difficult or impossible. Yet, because establishing communication is such an important goal, each patient showing willful modulation should be screened for candidacy for communication.

12.6 Patient-Centered Medical Decision-Making

The process of medical decision-making in patients with DoC has been the subject of much discussion [35, 49, 50]. Most often, the question has been framed as whether patients in irreversible VS should continue to receive life-sustaining therapy including hydration and nutrition by medical means or if they should be allowed to die by discontinuing these therapies [51–54]. This physician-centered locus of medical decision-making for patients with DoC is archaic and inappropriate because it wrongfully substitutes the values and preferences of the physician for those of the patient, thereby disenfranchising patients from their human right of self-determination.

It has been observed in this context that the appropriate locus of medical decision-making should be patient centered, i.e., the correct decision is that which follows and respects the patient's wishes and preferences [55]. They pointed out that "patient centeredness [is] an end in itself and a legitimate aspect of health care quality that must be measured and improved." Thus, physicians must devote effort to divine the wishes of the patient whose values and preferences should guide medical decision-making for their own treatment. The patient's preferences can be sought in prior executed advance directives for medical care and past statements made by the patient to relatives, friends, and physicians. The lawful surrogate decision-maker is authorized to make medical decisions for the patient in collaboration with the attending physician of record in a process of shared decision-making following their best understanding of the patient's wishes [56].

The very few VS or MCS patients who demonstrate willful modulation of brain activity by fMRI and have successfully achieved yes-no communication systems raise the question of the extent to which this capacity qualifies them to actively participate in their own medical decision-making. The closest analogy to this novel circumstance is the patient in a locked-in syndrome who retains normal cognition and awareness but is so profoundly paralyzed as to be incapable of communicating except for yes-no responses using a retained rudimentary voluntary movement, such as blinking or moving the eyes vertically. There has been much work accomplished to establish effective communication systems for locked-in patients using computers, lasers, and other technologies [57].

The essential question is whether the reliability of a binary yes-no response by fMRI achieves the threshold to validate a patient's consent or refusal of therapy. In ordinary clinical circumstances, physicians conduct discussions with patients or surrogates, explaining treatments and outcomes and answering questions in a consent process leading to their understanding and ability to reach a treatment decision. Most neurologists have concluded that because DoC patients demonstrating willful modulation, like those in a locked-in syndrome, can communicate only with rudimentary yes-no fMRI responses that categorically lack nuance and cannot convey understanding, these responses are too crude to achieve the criteria for a valid decision to continue or stop life-sustaining therapy. Therefore, surrogate decision-making is essential to make the patient's consent or refusal valid.

12.7 Medical Treatment

Medical treatment decisions for patients with DoC extend far beyond the vexing question of continuing life-sustaining medically supplied hydration and nutrition. Three classes of treatment have been impacted by innovations in functional neuroimaging: novel stimulatory protocols to attempt to improve responsiveness, neurorehabilitation, and palliative care.

Neurologists and physiatrists have attempted to stimulate patients in VS and MCS to try to improve their level of responsiveness using medications, environmental and physical stimuli, and more recently, electrical brain stimulation. A Cochrane systematic review of stimulation treatments in VS patients concluded that they are unsuccessful [58]. Amantadine is the exception: in a controlled trial, it was shown to improve the responsiveness of TBI patients with VS and MCS [59]. Patients in MCS are much more responsive to medications and other stimulation treatments; several treatments have shown benefit in improving their function [10]. Physicians should consider prescribing sequential trials of amantadine, levodopa, zolpidem [60], and other medications to stimulate patients with MCS in an attempt to improve their responsiveness [8].

A provocative report of deep brain stimulation of thalamic intralaminar nuclei inducing unequivocal improvement in the alertness, awareness, interaction, and speech in a chronic MCS patient has kindled interest in this technology as a treatment intervention in selected MCS patients [61]. MCS patients who might plausibly improve with deep brain electrical stimulation are those with intact medial thalamic neurons and their connections, as was true in one MCS patient selected for the study, because they retain the neuronal apparatus and connections to respond to stimulation [62]. The emerging technology of transcranial direct current stimulation of the left dorsolateral prefrontal cortex was recently reported to improve responsiveness in 30 MCS patients in a controlled, randomized, double-blind study, but the same treatment produced no beneficial effect in any of 25 VS patients [63].

Neurorehabilitation in a specialized unit is the optimal treatment for patients with disorders of consciousness after discharge from acute medical care [64]. Early skilled neurorehabilitative intervention has been shown to improve functional outcome in patients with DoC [65]. Functional neuroimaging may be helpful in these patients to guide the program of neurorehabilitation by clarifying their precise cognitive capacities and by identifying the cortical areas that have remained intact – though disconnected – that might offer opportunities for functional restoration by interventional neuromodulation [66].

Palliative care is an essential element of the treatment of patients with DoC [67]. This holistic approach to treatment encompasses patient- and family-centered care with consideration of emotional, psychological, social, and spiritual factors. Its intent is to maximize comfort and quality of life irrespective of the remaining quantity of life. Including the palliative care team in the management of DoC patients insures that optimal attempts will be made to communicate with the patient and to align the goals of therapy with those the patient would want.

12.8 Legal Issues

Functional neuroimaging in patients with DoC raises interesting legal questions, most of which exceed the scope of this chapter but have been analyzed elsewhere [68, 69]. A few of these legal issues impact medical practice on patients with DoC: informed consent for treatment, neurological diagnosis as it pertains to certain laws, the admissibility of fMRI findings and other scientific brain data into courtroom proceedings, and legal aspects of decisions to withhold or withdraw life-sustaining treatment.

The legal standards for informed consent for treatment vary among jurisdictions but generally require that a capacitated patient or the lawful surrogate decision-maker of an incapacitated patient consents for the medical plan once it has been explained by physicians and becomes understood by the surrogate in a noncoercive conversation. Patients with DoC are nearly always considered incapacitated, so the consent discussion is conducted with the patient's lawful surrogate. The finding of willful modulation of brain activity in a patient clinically diagnosed in VS or MCS does not itself render the patient capacitated to make a medical decision. But with the knowledge that the patient has some degree of awareness and comprehension, physicians have a duty also to explain the proposed treatment to the patient.

Functional neuroimaging findings can exert a legal impact on medical decisions in those jurisdictions that have statutory or case law governing the care of patients with specific DoC syndromes such as VS. Because of its unique features and analysis in highly publicized legal precedential cases, such as *Quinlan* and *Cruzan* in the United States and *Bland* in the United Kingdom [70], the VS has been written into some laws governing decisions to treat. For a VS patient residing in such a jurisdiction, if the fMRI showed willful modulation of brain activity resulting in the diagnosis being changed from VS to MCS, that fact plausibly could impact on whether the law continues to apply to them. For example, in the United States, the highly publicized *Wendland* legal ruling that barred the family surrogate decision-maker from authorizing discontinuation of his life-sustaining treatment depended on the fact that the patient was in MCS, not in VS [71].

The issue of whether brain fMRI and other "inferential" scientific data concerning a person's state of consciousness or behavior can be admitted in a court proceeding is an area of emerging law. Although not primarily a clinical issue involving medical practice, lawyers have attempted to admit brain fMRI data (e.g., of abnormal findings correlated with antisocial behavior) as exculpatory evidence in criminal trials with varying degrees of success [72]. It is conceivable that such data might also be admitted to civil cases that would require physician testimony and thereby impact medical practice.

Decisions to withdraw or withhold life-sustaining therapy on patients with DoC have obvious legal restrictions and implications. Because they vary greatly among jurisdictions, physicians caring for DoC patients should be knowledgeable about the applicable laws in force where they practice. Functional neuroimaging findings could impact on this question if they changed the diagnosis from VS to MCS. The medical, ethical, and legal issues involved in these decisions in patients with DoC and other neurological disorders were reviewed recently [73]. The specific laws in the United States governing discontinuation of medically administered hydration and nutrition from VS patients also have been reviewed [74].

12.9 Future Directions

More systematic studies of patients with DoC are needed because most of the existing data consist of single case reports or small series. Patients with DoC comprise an "orphan" disease category in which funding to conduct such studies is limited.

Unless a pharmaceutical or medical device intervention was found that could be applied widely and thereby be profitable, this group of patients is of relatively little interest to industry research sponsors. Private funding agencies and philanthropies constitute the main source of current research support.

Further studies can help to better understand the anatomy and physiology of consciousness, the process by which diseases and trauma impair it, and the interventions might improve or palliate DoC patients. It seems likely that the clinical syndromes, such as VS and MCS, will be redrawn along more physiological lines with less reliance on clinical syndromes for diagnosis. The new diagnostic categories will incorporate findings of functional neuroimaging and brain electrical potential studies. The new nosology will permit a more individualized diagnostic approach that accommodates the specific areas of brain damage and resulting impairments of function.

A great advance in the care of patients with DoC will be the further enhancement of communication technologies using fMRI, FDG-PET, or EEG data. These technologies will reduce the suffering of patients with DoC by allowing them to explain their symptoms, fears, and hopes while allowing them to participate in rehabilitation, palliation, and medical treatment decisions that affect them. But before this desirable outcome can occur, the fMRI, FDG-PET, and other experimental paradigms will need to be validated, standardized, and commercialized to permit their more widespread dissemination into medical practice [75].

References

1. Fernandez-Espejo D, Owen AM (2013) Detecting awareness after severe brain injury. Nat Rev Neurosci 14:1–9
2. Giacino JT, Fins JJ, Laureys S, Schiff ND (2014) Disorders of consciousness after acquired brain injury: the state of the science. Nat Rev Neurol 10:99–114
3. Gosseries O, Di H, Laureys S, Boly M (2014) Measuring consciousness in severely damaged brains. Annu Rev Neurosci 37:457–478
4. Fins JJ, Illes J, Bernat JL, Hirsch J, Laureys S, Murphy E (2008) Neuroimaging and disorders of consciousness: envisioning an ethical research agenda. Am J Bioeth 8(9):3–12
5. Fins JJ (2003) Constructing an ethical stereotaxy for severe brain injury: balancing risks, benefits and access. Nat Rev Neurosci 4:323–327
6. Illes J, Lau PW, Giacino JT (2008) Neuroimaging, impaired states of consciousness, and public outreach. Nat Clin Pract Neurol 4:542–543
7. Graham M, Weijer C, Peterson A et al (2015) Acknowledging awareness: informing families of individual research results for patients in the vegetative state. J Med Ethics 41:534–538
8. Bernat JL (2009) Ethical issues in the treatment of severe brain injury: the impact of new technologies. Ann N Y Acad Sci 1157:117–130
9. Bernat JL (2010) Current controversies in states of chronic unconsciousness. Neurol Clin Pract 75(Suppl 1):S33–S38
10. Bernat JL (2009) Chronic consciousness disorders. Annu Rev Med 60:381–392
11. Bruno M-A, Laureys S, Demertzi A (2013) Coma and disorders of consciousness. Handb Clin Neurol 118:205–213
12. Owen AM (2008) Disorders of consciousness. Ann N Y Acad Sci 1124:225–238

13. Owen AM, Coleman MR, Boly M, Davis MH, Laureys S, Pickard JD (2006) Detecting awareness in the vegetative state. Science 313:1402
14. Bardin JC, Fins JJ, Katz DI et al (2011) Dissociations between behavioural and functional magnetic imaging-based evaluations of cognitive functioning after brain injury. Brain 134:769–782
15. Bardin JC, Schiff ND, Voss HU (2012) Pattern classification of volitional magnetic resonance imaging responses in patients with severe brain injury. Arch Neurol 69:176–181
16. Monti MM, Vanhaudenhuyse A, Coleman MR et al (2010) Willful modulation of brain activity in disorders of consciousness. N Engl J Med 362:579–589
17. Rosanova M, Gosseries O, Casarotto S et al (2012) Recovery of cortical effective connectivity and recovery of consciousness in vegetative patients. Brain 135:1308–1320
18. Monti MM, Rosenberg M, Finoia P, Kamau E, Pickard JD, Owen AM (2015) Thalamo-frontal connectivity mediates top-down cognitive functions in disorders of consciousness. Neurology 84:167–173
19. Celesia GG (2013) Conscious awareness in patients in vegetative states: myth or reality? Curr Neurol Neurosci Rep 13:395–397
20. Ropper AH (2010) *Cogito ergum sum* by MRI. N Engl J Med 362:648–649
21. Billings JA, Churchill LR, Payne R (2010) Severe brain injury and the subjective life. Hastings Cent Rep 40(3):17–21
22. Fins JJ (2012) Wait, wait …don't tell me: tuning in the injured brain. Arch Neurol 69:158–160
23. Cruse D, Chennu S, Chatelle C et al (2012) Relationship between etiology and covert cognition in the minimally conscious state. Neurology 78:816–822
24. Fingelkurts AA, Fingelkurts AA, Bagnato S, Boccagni C, Galardi G (2012) EEG oscillatory states as neuro-phenomenology of consciousness as revealed from patients in vegetative and minimally conscious states. Conscious Cogn 21:149–169
25. Forgacs PB, Conte MM, Fridman EA, Voss HU, Victor JD, Schiff ND (2014) Preservation of electroencephalographic organization in patients with impaired consciousness and imaging-based evidence of command-following. Ann Neurol 76:869–879
26. Faugeras F, Rohaut B, Weiss N et al (2011) Probing consciousness with event-related potentials in the vegetative state. Neurology 77:264–268
27. Luauté J, Cotton F, Lemaire J-J et al (2012) Let live or let die after traumatic coma: scrutinizing somatosensory evoked potentials. Neurol Clin Pract 2(1):24–32
28. Avidan MS, Zhang L, Burnside BA et al (2008) Anesthesia awareness and the bispectral index. N Engl J Med 358:1097–1108
29. Cruse D, Chennu S, Bekinschtein TA et al (2011) Bedside detection of awareness in the vegetative state: a cohort study. Lancet 378:2088–2094
30. Goldfine AM, Bardin JC, Noirhomme Q, Fins JJ, Schiff ND, Victor JD (2013) Reanalysis of "bedside detection of awareness in the vegetative state: a cohort study". Lancet 381:289–291, and author's reply (2013). Lancet 381:291–292
31. Multi-Society Task Force on PVS (1994) Medical aspects of the persistent vegetative state: parts I and II. N Engl J Med 330:1499–1508, 1572–1579
32. Bernat JL (2013) Increasing awareness in unawareness. JAMA Neurol 70:1231–1232
33. Schnakers C, Vanhaudenhuyse A, Giacino J et al (2009) Diagnostic accuracy of the vegetative and minimally conscious state: clinical consensus versus standardized neurobehavioral assessment. BMC Neurol 9:35
34. Giacino JT, Kalmar K, Whyte J (2004) The JFK coma recovery scale-revised: measurement characteristics and diagnostic utility. Arch Phys Med Rehabil 85:2020–2029
35. Racine E, Rodrigue C, Bernat JL, Riopelle R, Shemie SD (2010) Observations on the ethical and social aspects of disorders of consciousness. Can J Neurol Sci 37:758–768
36. Coleman MR, Davis MH, Rodd JM et al (2009) Towards the routine use of brain imaging to aid the clinical diagnosis of disorders of consciousness. Brain 132:2541–2552
37. Stender J, Gosseries O, Bruno M-A, Charland-Verville V, Vanhaudenhuyse A, Demertzi A et al (2014) Diagnostic precision of PET imaging and functional MRI in disorders of consciousness: a clinical validation study. Lancet 384:514–522

38. Estraneo A, Moretta P, Loreto V et al (2010) Late recovery after traumatic, anoxic, or hemorrhagic long-lasting vegetative state. Neurology 75:239–245
39. Estraneo A, Moretta P, Loreto V et al (2013) Predictors of recovery of responsiveness in prolonged anoxic vegetative state. Neurology 80:464–470
40. Luauté J, Maucort-Boulch D, Tell L et al (2010) Long-term outcomes of chronically minimally conscious and vegetative states. Neurology 75:246–252
41. Steppacher I, Kaps M, Kissler J (2014) Will time heal? A long-term follow-up of severe disorders of consciousness. Ann Clin Transl Neurol 1:401–408
42. Di HB, Yu SM, Weng XC et al (2007) Cerebral response to patient's own name in the vegetative and minimally conscious states. Neurology 68:895–899
43. John ER, Halper JP, Lowe RS III, Merkin H, Defina P, Prichep LS (2011) Source imaging of QEEG as a method to detect awareness in a person in a vegetative state. Brain Inj 25:426–432
44. Katz DI, Polyak M, Coughlan D et al (2009) Natural history of recovery from brain injury after prolonged disorders of consciousness: outcome of patients admitted to inpatient rehabilitation with 2-4 year follow-up. Prog Brain Res 177:73–88
45. Whyte J, Gosseries O, Chervoneva I et al (2009) Predictors of short-term outcome in brain-injured patients with disorders of consciousness. Prog Brain Res 177:63–72
46. Voss HU, Uluğ AM, Dyke JP et al (2006) Possible axonal regrowth in late recovery from the minimally conscious state. J Clin Invest 116:2005–2011
47. Nakase-Richardson R, Yablon SA, Sherer M, Nick TG, Evans CC (2009) Emergence from minimally conscious state: insights from evaluation of posttraumatic coma. Neurology 73:1120–1126
48. Naci L, Owen AM (2013) Making every word count for nonresponsive patients. JAMA Neurol 70:1235–1241
49. American Academy of Neurology Quality Standards Subcommittee (1995) Practice parameters: assessment and management of patients in the persistent vegetative state (summary statement). Neurology 45:1015–1018
50. National Health and Medical Research Council (2008) Ethical guidelines for the care of people in post-coma unresponsiveness. Australian National Health and Medical Research Council: Canberra, Australia. [http://www.nhmrc.gov.au/guidelines-publications/e81-e82]
51. Constable C (2012) Withdrawal of artificial nutrition and hydration for patients in a permanent vegetative state: changing tack. Bioethics 26(3):157–163
52. Gipson J, Kahane G, Savulescu J (2014) Attitudes of lay people to withdrawal of treatment in brain damaged patients. Neuroethics 7:109
53. Varelius J (2013) Pascal's Wager and deciding about life-sustaining treatment of patients in persistent vegetative state. Neuroethics 6:277–285
54. Wilkinson DJ, Kahane G, Horne M, Savulescu J (2009) Functional neuroimaging and withdrawal of life-sustaining therapy from vegetative patients. J Med Ethics 35:508–511
55. Holloway RG, Quill TE (2010) Treatment decisions after brain injury – tensions among quality, preference, and cost. N Engl J Med 362:1757–1759
56. King JS, Moulton BW (2006) Rethinking informed consent: the case for shared medical decision-making. Am J Law Med 32:429–501
57. Laureys S, Pellas F, Van Eeckhout P et al (2009) The locked-in syndrome: what it is like to be conscious but paralyzed and voiceless. Prog Brain Res 150:495–511
58. Lombardi F, Taricco M, De Tanti A, Telaro E, Liberati A (2002) Sensory stimulation of brain-injured individuals in coma or vegetative state: results of a Cochrane systematic review. Clin Rehabil 16:464–472
59. Giacino JT, Whyte J, Bagiella E et al (2012) Placebo-controlled trial of amantadine for severe traumatic brain injury. N Engl J Med 366:819–826
60. Brefel-Courbon C, Payoux P, Ory F et al (2007) Clinical and imaging evidence of zolpidem effect in hypoxic encephalopathy. Ann Neurol 62:102–105
61. Schiff ND, Giacino JT, Fins JJ (2009) Deep brain stimulation, neuroethics, and the minimally conscious state: moving beyond proof of principle. Arch Neurol 66:697–702

62. Giacino J, Fins JJ, Machado A, Schiff ND (2012) Central thalamic deep brain stimulation to promote recovery from chronic posttraumatic minimally conscious state: challenges and opportunities. Neuromodulation 15:339–349
63. Thibaut A, Bruno M-A, Ledoux D, Demertzi A, Laureys S (2014) tDCS in patients with disorders of consciousness: sham-controlled randomized double-blind study. Neurology 82:1112–1118
64. Giacino JT, Katz DI, Whyte J (2013) Neurorehabilitation in disorders of consciousness. Semin Neurol 33:142–156
65. Eilander JJ, Wijnen VJ, Scheirs JG, de Kort PL, Prevo AJ (2005) Children and young adults in a prolonged unconscious state due to severe brain injury: outcome after early intensive neurorehabilitation programme. Brain Inj 19:425–436
66. Giacino JT, Hirsch J, Schiff N, Laureys S (2006) Functional neuroimaging applications for assessment and rehabilitation planning in patients with disorders of consciousness. Arch Phys Med Rehabil 87(12 Suppl 2):S67–S76
67. Fins JJ (2005) Clinical pragmatism and the care of brain-damaged patients: toward a palliative neuroethics for disorders of consciousness. Prog Brain Res 150:569–586
68. Bressman JO, Reidler JS (2010) Willful modulation of brain activity in disorders of consciousness: legal and ethical ramifications. J Law Med Ethics 38:713–716
69. Fisher CE, Appelbaum PS (2010) Diagnosing consciousness: neuroimaging, law, and the vegetative state. J Law Med Ethics 38:374–385
70. Bernat JL (2008) Ethical issues in neurology, 3rd edn. Lippincott Williams & Wilkins, Philadelphia
71. Lo B, Dornbrand L, Wolf LE, Groman M (2002) The Wendland case –withdrawing life support from incompetent patients who are not terminally ill. N Engl J Med 346:1489–1493
72. Yaffe G (2013) Neurologic disorder and criminal responsibility. Handb Clin Neurol 118:345–356
73. Cochrane TI (2013) Withdrawing and withholding life-sustaining treatment. Handb Clin Neurol 118:147–153
74. Larriviere D, Bonnie RJ (2006) Terminating artificial nutrition and hydration in persistent vegetative state patients: current and proposed state laws. Neurology 66:1624–1628
75. Sleigh J, Warnaby CE (2014) Functional brain imaging: gatecrashing the clinical party? Lancet 384:476–477

Chapter 13
Vegetative State Two Decades After the Multi-Society Task Force (MSTF) Report

Gastone G. Celesia

Abstract Two decades after the MSTF report:

- The classification of patients with DoC has improved due to the use of stringent protocols.
- Clinical/behavior evaluation remains the golden standard in determining VS/MCS.
- Custodial care of VS/MCS patients has improved.
- Neuroimaging has shown that the vegetative brain is more complex than previously envisioned and may retain some capacity for higher levels of neuro-functioning.
- The prognostic indicators of outcome remain etiology, age, and duration of VS/MCS.
- No treatment for VS/MCS has emerged, and the patient's outcome remains poor.
- The cost/benefit analysis of VS shows a severe burden to family and society and must be considered in the evaluation of nation health-care needs.
- The issue of "consciousness" and "perception of pain" remains an insolvable conundrum.

Keywords Vegetative state • Clinical evaluation • Neuroimaging • Prognosis • Treatment

13.1 Introduction

Vegetative state (or unresponsive wakefulness syndrome) (VS) became a health issue in the last four decades as a result of aggressive medical intervention attempting to "…rescue from life-threatening events…" patients by resuscitation and intensive care [1]. VS is the unintended consequence of these efforts. Two

G.G. Celesia, MD, MSc
Department of Neurology, Loyola University of Chicago, Maywood, IL, USA

Chicago Council on Science and Technology, Chicago, IL, USA
e-mail: g.celesia@comcast.net

M.M. Monti, W.G. Sannita (eds.), *Brain Function and Responsiveness in Disorders of Consciousness*, DOI 10.1007/978-3-319-21425-2_13

milestones in our understanding of VS have been the publication of Jennett and Plum in 1972 [2] that coined the term persistent vegetative state (PVS) and the Multi-Society Task Force (MSTF) that in 1994 [3] came to a consensus definition of VS ad its prognosis.

MSTF defined VS as "...an irregular but cyclic state of circadian sleeping and waking unaccompanied by any behaviorally detectable expression of self-awareness, specific recognition of external stimuli, or consistent evidence of attention or intention or learned response..." [3]. There is now a controversy about the definition of VS, the validity of the prognostic indicators, and the use of clinical observations of behavior as the sole criterion for diagnosis. At the Aspen Neurobehavioral Conference, the condition of minimally conscious state (MCS) was defined, showing that some patients believed to be in VS were in fact minimally conscious [4]. What has been the significant, if any, breakthrough in the diagnosis and management of VS patients? Is there any evidence that the present therapeutic intervention has improved the patient outcome? Is there any new evidence that we should change or amend the recommendations of the MSTF report?

13.2 What Is New in Diagnosis and Management?

The diagnoses of VS/MCS have remained difficult. Misdiagnosis was reported to vary from 37 to 43 % [5–7] and "...despite the importance of diagnostic accuracy, the rate of misdiagnosis of VS has not substantially changed in the past 15 years..." [7]. These data have spurred efforts to improve the diagnostic accuracy by using standardized protocols and ancillary tests [4, 8–14]. The Glasgow Coma Scale, Full Outline of Unresponsiveness scale, Wessex Head Injury Matrix, Sensory Modality Assessment and Rehabilitation Technique, Los Amigos Cognitive Scale-R, and Coma Recovery Scale-Revised have been proposed to assess the level of consciousness [15–18]. Among the ancillary tests are neurophysiological (EEG, ERP) and neuroimaging (PET, SPECT, fMRI) tests, although their diagnostic accuracy remains controversial [19, 20]. Gibson et al. [21] advocate the use of multiple tasks and neuroimaging modalities to assess "covert awareness," methodology useful as a research tool, but not practical in the clinical setting. Clinical/behavioral evaluation remains the golden standard in determining VS/MCS [22].

The patient's consciousness status varies in time [23]; thus at one particular observation, the subject may be in MCS while in the next observation the same subject may be in VS. VS and MCS represent continuous bidirectional states [23] and may indicate a gradual recovery from coma or a regression toward the end of life (Fig. 13.1). Multiple behavior observations and the utilization of standard protocols have improved diagnosis, although the expectation of 100 % accuracy remains elusive and unattainable.

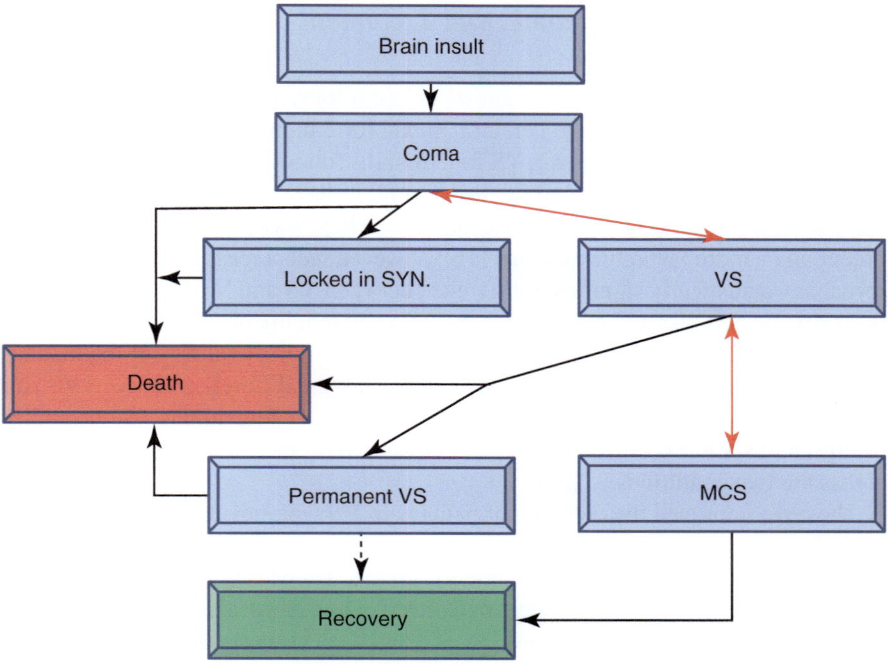

Fig. 13.1 Flow chart of patients in coma and VS. The *red arrows* indicate that the process is bidirectional. The *dashed arrow* indicates that this is a rare occurrence. Permanent VS is defined more than 3 months if nontraumatic and more than 1 year if traumatic. *SYN = syndrome*, *VS = Vegetative State*, and *MCS = minimally conscious state* (modified from Celesia [55])

13.2.1 Incidence and Prevalence of VS/MCS

Incidence depends "…heavily upon the period post-incident…" when VS is considered and is subject to a "…variety of sources of error…" [25]. The incidence of VS continuing for at least 6 months varies from 0.5 to 25 per 100,000 of population [24, 25]. The prevalence varies among reports and countries [25–27]. The variability is related to the "…absence of a gold standard for the diagnosis of VS…" [26], whether cases of VS due to degenerative nervous system diseases are included or excluded and the "…quality and availability of emergency and intensive care services…" [26]. None of these factors will explain the lower prevalence observed in Denmark and the Netherlands compared to the other European nations that is related to the moral, religious, and societal "…end-of-life decisions in the intensive care unit, on hospital wards…," and long-term care facilities [26, 27]. Here we are facing with the dichotomy of two divergent philosophies: the right to a dignified death [28, 29] and the sanctity of life (pro vita) [30, 31]. This then is no longer a medical issue but a political/religious/moral issue.

13.2.2 Prognosis, Outcome, and Treatment

The MSTF major indicators for prognosis were etiology, age, and length of VS [32]. An elderly person in VS due to cerebral anoxia for 3 months has almost no chance of recovery, whereas a teenager in VS for 3 months following a head trauma still has a greater than 24 % chance of some recovery [1, 4, 9, 23, 32, 33]. These remain as solid prognostic criteria. The identification of MCS was greeted as a state with potentially a more favorable outcome [16]; however, recent reassessment "...did not reveal any statistically significant difference between VS and MCS...." Overall, "... there was no combination of variables that allowed reliably discriminating between VS and MCS..." and "...results cast doubt on the empirical validity of the distinction between VS and MCS..." [34]. Whether or not the difference between VS and MCS is valid, "...patients in VS and MCS have similar functioning and disability profiles and similar needs thus levels of care and assistance should not be different across the two conditions..." [35].

Have we improved the outcome of patients in VS with novel interventions? The answer is negative. Although modern technology and supportive care have improved the custodial/palliative care of VS patients, no medical or surgical intervention has shown any influence on the patient's outcome. Claims of positive effects have been made using a variety of medications [36–40] and deep brain stimulation [41]. These reports have flaws in their design, they lack "...double-blind placebo controlled studies..." and defined inclusion and exclusion criteria, and they use "...ill-defined recovery criteria..." [42], rendering their claims unreliable [43]. When double-blind placebo-controlled studies were done, the data showed no effect of zolpidem or amantadine on the outcome of VS/MCS [44, 45].

13.3 Neuroimaging in VS/MCS

Neuroimaging tests (fMRI, PET) have been applied to evaluate the cerebral activity of VS/MCS [46–72], with contributions in research, diagnosis and prognosis, and assessment of "conscious awareness."

13.3.1 Research

fMRI/PET have shed new light in our understanding of consciousness and have shown that the vegetative brain is more complex than previously envisioned [46–69]. The activation of cerebral structures to complex language stimuli or to mental imagery suggests that the VS brain may retain some capacity for higher levels of neurofunctioning [48–52, 61].

13.3.2 Diagnosis and Prognosis

fMRI and PET improve the diagnostic accuracy of VS and help in differentiating VS from MCS [46–53, 68]. Hannawi et al. [47] reported reduced activity in the cingulate gyrus in all VS and MCS subjects and noted that in VS, the reduction affected bilateral medial dorsal thalamic nuclei, while in MCS, the reduction was limited to the right medial dorsal nucleus, thus separating the two states [47]. fMRI/PET may also be helpful in the prognostication of recovery [18, 49, 56, 57]. The presence of activation of primary sensory cortices, higher-order associative areas, and subcortical areas to either mental imagery tasks or high-level language stimuli is often associated with some recovery.

13.3.3 VS and "Conscious Awareness"

Whether fMRI/PET can establish consciousness and evaluate if the patient feels pain remain controversial. Neuroimaging responses to complex language stimuli or to mental imagery in *some subset* of VS patients are similar to the response of normal controls [13, 50–52, 56–58, 64]. Does this correlation indicate "conscious awareness"? Similarly, neuroimaging activation of cerebral regions to painful stimuli implies pain perception [54, 66–69].

There is a consensus that the utilization of complex language stimuli activates in *some* VS/MCS patients' brain regions is associated with language processing (bilateral superior temporal planes, superior temporal gyri, inferior frontal gyri, and left posterior inferior temporal cortex) [50, 55–59, 65]. Similarly, the use of mental imagery paradigms in *some* VS patients activates brain structures that are similar to the activation in normal controls [51–53]. There are two possible interpretations of these data. Proponents of "functional neuroimaging" believe that brain activation demonstrates "neuronal correlates of speech comprehension" [56], a state of consciousness [50] and a retained capacity for "covert cognition." They conclude that neuroimaging activation is a signature of some conscious processing [13, 21, 48–52, 56–60, 65, 69, 71, 72]. The alternative interpretation is based on the demonstration of reduced and disrupted brain activation and on clinical observations. Hannawi et al. [47] in a meta-analysis of 13 studies (272 patients, 259 healthy controls) noted "…consistently reduced activity in patients with DoC in bilateral medial dorsal nucleus of the thalamus, left cingulate, posterior cingulate, precuneus, and middle frontal and medial temporal gyri". Stender [70] showed a decreased of global cerebral metabolism (indicating decreased neuronal activity) in VS patients compared to normal controls (Fig. 13.2). Many studies have shown that brain connectivity is disrupted in VS/MCS [46, 50, 64, 68, 79]. Dehaene and Changeux [73] suggest that conscious awareness occurs "when incoming information is made globally available to multiple brain systems through a network of neurons with long-range axons

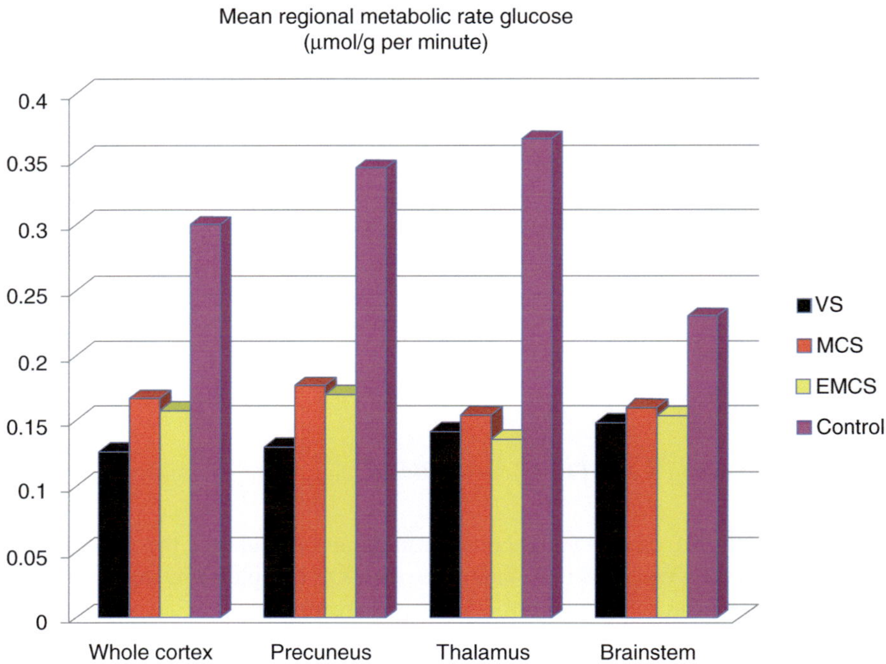

Fig. 13.2 Mean regional cerebral metabolic rates of glucose (μmol/g per minute). Note the reduced regional cerebral metabolic rate from the whole cortex, precuneus, thalamus, and brainstem in impaired consciousness compared to controls. *EMCS* emergence from minimally conscious state, *MCS* minimally conscious state, *VS* vegetative state (data plotted from Stender et al. [70])

densely distributed in prefrontal, parieto-temporal, and cingulate cortices." It is therefore conceivable that fMRI activation to language and imagery in some VS subjects simply reflects the preservation of some modules, while these modules have disrupted connectivity and lack the integrative network processes necessary for normal cognitive functions [55, 74].

Furthermore, 46/121 (38 %) patients in MCS who by definition are supposed to have some minimal consciousness showed no response or limited activation on fMRI, suggesting absence of any awareness [55]. This dissociation between behavioral observation and neuroimaging should set a cautionary note on any of our interpretations. The other issue closely correlated to conscious awareness is the issue of pain perception. Are behavioral responses to pain observed in VS patients (crying, withdrawing from the stimulus, and changes in heart rate) an expression of emotional primordial reflexes or indication of some awareness? Can the information gathered by neuroimaging answer this question? 60 % of VS and 19 % of MCS showed none or minimal cerebral activation to nociceptive stimuli. The remainder of VS/MCS patients showed fMRI/PET complex cerebral activation to nociceptive stimuli, demonstrating some neuronal processing of pain [66–69]. Whether "…are images of increased blood oxygen level–dependent signal (fMRI) or oxygen/glucose uptake (PET) equivalent to conscious perception, or indicative of awareness, or are they simply document-retained modular function in the absence of the integrative processes necessary for consciousness…" remains an open question. VS subjects are unable to communicate their feelings and wishes and to manifest volition. Present technology no matter how sophisticated cannot read or interpret a person's mind. Are we willing to accept neuroimaging tests as surrogates for communication among us and VS subjects? Controversial interpretation of data does not offer sufficient evidence to proxy a machine as a "sentient communicator"; thus, the issue of residual consciousness and pain perception in VS remains an unresolved Gordian knot [54, 55, 74]. The lay press has become concerned by some of our claims and editorializes against "people who take pretty brain-scan images and claim they can use them to predict what product somebody will buy, what party they will vote for, whether they are lying or whether a criminal should be held responsible for his crime" [75] or what they think [76]. David Brooks of the New York Times said:"…an important task these days is to harvest the exciting gains made by science and data while understanding the limits of science and data. The next time somebody tells you what a brain scan says, be a little skeptical…" [75].

13.4 VS/MCS Ontological Status

The "question of the patients' ontological status" [77] has remained the same in the last 20 years. VS patients according to Winkler "…are periodically awake, and their bodies breathe and digest on their own. These traits bespeak life. Yet they are not conscious and never will be: subjectively, this is death…" [78]. VS patients are non-sentient, "nonautonomous," because "they lack the capacity to make decisions" [2, 3, 78–81]; others have described VS as a limbo and a state worse than death [82–84]. MCS patients by definition are minimally conscious, however are unable to communicate or express their wishes; thus, they are also "nonautonomous." The advocates of fMRI to determine the presence of "covert consciousness" insist that patients with cerebral activation to language or imagery are somewhat sentient. However, "…it is not always reasonable to assume that because a brain area is active, a commonly associated cognitive function is engaged...." This process of inferring from the presence of brain activation the presence of a cognitive function is *reverse inference* and a *logical error* [85, 86].

All VS/MCS patients will die if left without a sustained level of chronic care (including feeding, hydration, nursing care, etc.). "If the only achievement of aggressive care is the maintenance of a non-sentient body, is this choice morally justified? Are we refusing to face death by maintaining alive bodies without a mind? I do not have, and I believe nobody has, absolute answers to these vexing questions" [81].

Who should decide the fate of VS/MCS subjects, whether to provide continued care or withdrawal nutrition and hydration? These are the possible candidates:

- The individual in VS/MCS (if his/her wishes prior to sickness are known or via advance directive)
- The next of keen/family/surrogate
- The health-care provider (physicians, administrators)
- A machine (fMRI/PET, EEG/ERP)
- Politicians
- The court via legal intervention
- Religious leaders

In a perfect world, the right of the individual or his/her surrogate should prevail. Unfortunately, health-care providers, politicians, religious leaders, advocates of neuroimaging, and the courts have sometimes intervened, all in the name of the "patient's best interest" [28–31, 77–91]. The possibility that the decision is made by a machine, politicians, religious leaders or the courts raises the specter of *Big Brother* interfering with the independent constitutional right of the individual.

The legal ramifications on the decision of the patient fate vary from country to country [85, 88, 92, 93]. The introduction of fMRI/PET in some legal cases has resulted in confusion "…in the application of neuroimaging to legal tests of consciousness…" [85]. The dichotomy of two divergent philosophies – sanctity of life/inestimable value of life itself and the right of refusing treatment/the right of a dignified death – has spilled into the courts [85, 92]. The wishes of the patient and of the family should be paramount in the decision to continue or withdraw care. The patient's advance directive should not be ignored by the health-care staff or by government officials [94]. Conflicts may arise between the "do everything" and the "stop all care" groups. The differing opinions may involve physicians, family, and hospital administrators, with potential conflicts within and between groups. Too often, the fate of VS patients has become contentious [85, 92]. The administration of life-prolonging treatments is at time continued long "after their families have come to believe that the patient would rather be dead" [95]. Kitzinger and Kitzinger [95] suggest that medical ethicists concerned about the rights of people in VS/MCS need to take this empirical data into account in seeking to apply ethical theories to medicolegal realities.

13.5 Cost/Benefit Analysis

The health-care cost and the economic and family burden of caring for VS/MCS patients are seldom discussed.

13.5.1 Caregivers' Burden

The care of VS/MCS is affecting their family [96–98]. The distress and burden are "high for all caregivers" [96]. Leonardi et al. [97] suggest that there is a need for comprehensive support strategies for caregivers to "diminish the level of burden." Caregivers often report a loss of income estimated to be "less than 17,000 euros per year" [97].

13.5.2 Health-Care Cost

The health-care cost of caring for VS/MCS patients needs to be evaluated particularly in the present era of limited resources. Former Colorado Gov. Richard Lamm back in 1987 framed the issue succinctly: "We refuse to even discuss the reality that in an aging society with exploding numbers of elderly and exponentially rising health-care costs and myriad other unmet needs, the level of treatment for a 90-year-old might be different from that for a 9-year-old. Should we not put our limited dollars into more quality of a healthy life rather than more quantity of a sick life--more to the recoverable young rather than the terminally ill?" [99]. The cost of care of a woman that survived for 17 years in VS was estimated in 1981 at $6,104,590 ($ 61,028 per year) [100]. Kaufman and Lipton [101] obtained in 1992 the hospital bill of 13 patients in VS. Their mean hospital length of stay was 197 days and the averaged cost was $170,000. According to the US government data, the average costs for long-term care in 2010 were $205 per day or $74,820 per year [102]. The prevalence of VS in adults in the United States is estimated between 4 and 16.8/100,000 [25]. The US population in 2014 was 318,857,056; then the estimated number of VS/MCS patients is between 12,754 and 51,017 with a cost to the health care between one to four billion dollars/year. The MSTF [3, 32] similarly estimated the total annual cost for caring of VS patients in the United States to be between $1 and $7 billion. This is a huge burden to the health budget that cannot be overlooked. It may be time to set aside preconceived ideas and look pragmatically to the issues of VS in the contest of all others health needs and ask ourselves if the cost/benefit ratio is too much into the negative.

13.6 Conclusions

The use of stringent protocols has improved the classification of patients with disturbances of consciousness. The prognostic predictors of outcome remain: etiology, age, and duration of VS/MCS. Care of patients in VS has improved as facilities for prolonged care are available in most countries; however, there is still no treatment for VS/MCS and their outcome remains poor. fMRI and electrophysiology studies have not answered the pivotal questions of what interventions may benefit the patients and/or determine what society should recommend for these patients.

If we cannot offer any treatment but only custodial care, who is to decide the fate of these patients? This is no longer a medical decision; society must discuss these issues and based on moral/religious, cost/benefit analysis, and resource availability, determine what should be done. As society varies in different countries, we should not expect nor demand a solution fitting all; there are no wrong or right ways, but diverse solutions. The physician and neuroscientist's role is not to pontificate and decide but to offer unbiased data during the society deliberations and a supportive role whatever decision is finalized.

References

1. Jennett B (2005) Thirty years of the vegetative state: clinical, ethical and legal problems. Prog Brain Res 150:537–543
2. Jennett B, Plum F (1972) Persistent vegetative state after brain damage: a syndrome in search of a name. Lancet 1:734–737
3. Medical aspects of the persistent vegetative state [1]. The Multi-Society Task Force on PVS (1994) N Engl J Med 330:1499–1508
4. Giacino JT, Zasler ND, Katz DI, Kelly JP, Rosenberg JH, Filley CM (1997) Development of practice guidelines for assessment and management of the vegetative and minimally conscious states. J Head Trauma Rehabil 12:79–89
5. Childs NL, Mercer WN, Childs HW (1993) Accuracy of diagnosis of persistent vegetative state. Neurology 43(8):1465–1467
6. Andrews K, Murphy L, Munday R, Littlewood C (1996) Misdiagnosis of the vegetative state: retrospective study in a rehabilitation unit. BMJ 313(7048):13–16
7. Schnakers C, Vanhaudenhuyse A, Giacino J, Ventura M, Boly M, Majerus S, Moonen G, Laureys S (2009) Diagnostic accuracy of the vegetative and minimally conscious state: clinical consensus versus standardized neurobehavioral assessment. BMC Neurol 9(1):35
8. Fingelkurts AA, Bagnato S, Boccagni C, Galardi G (2013) Prognostic value of resting-state electroencephalography structure in disentangling vegetative and minimally conscious states: a preliminary study. Neurorehabil Neural Repair 27(4):345–354
9. Estraneo A, Moretta P, Loreto V, Lanzillo B, Cozzolino A, Saltalamacchia A, Lullo F, Santoro L, Trojano L (2013) Predictors of recovery of responsiveness in prolonged anoxic vegetative state. Neurology 80(5):464–470
10. Cruse D, Chennu S, Fernandez-Espejo D, Payne WL, Young GB, Owen AM (2012) Detecting awareness in the vegetative state: electroencephalographic evidence for attempted movements to command. PLoS One 7(11):e49933
11. Heiss WD (2011) PET in coma and in vegetative state. Eur J Neurol 19(2):207–211
12. Fernandez-Espejo D, Bekinschtein T, Monti MM, Pickard JD, Junque C, Coleman MR, Owen AM (2011) Diffusion weighted imaging distinguishes the vegetative state from the minimally conscious state. Neuroimage 54(1):103–112
13. Bekinschtein TA, Manes FF, Villarreal M, Owen AM, Della-Maggiore V (2011) Functional imaging reveals movement preparatory activity in the vegetative state. Front Hum Neurosci 5:5
14. Bardin J, Fins JJ, Katz DI, Hersh J, Heier LA, Tabelow K, Dyke JP, Ballon DJ, Schiff ND, Voss H (2011) Dissociations between behavioural and functional magnetic resonance imaging-based evaluations of cognitive function after brain injury. Brain 134:769–782
15. Schnakers C (2012) Clinical assessment of patients with disorders of consciousness. Arch Ital Biol 150(2–3):36–43
16. Giacino J, Ashwal S, Childs N, Cranford R, Jennett B, Katz DI, Kelly JP, Rosenberg JH, Whyte J, Zafonte RD (2002) The minimally conscious state: definition and diagnostic criteria. Neurology 58:349–353
17. Guldenmund P, Stender J, Heine L, Laureys S (2012) Mindsight: diagnostics in disorders of consciousness. Crit Care Res Pract. 13 p Article ID 624724. http://dx.doi.org/10.1155/2012/624724
18. Bosco A, Lancioni GE, Olivetti Belardinelli M, Singh NN, O'Reilly MF, Sigafoos J (2010) Vegetative state: efforts to curb misdiagnosis. Cogn Process 11(1):87–90
19. Peterson A, Cruse D, Nacia L, Weijera C, Owen AM (2015) Risk, diagnostic error, and the clinical science of consciousness. Neuroimage Clin 7:588–597
20. Goldfine AM, Bardin JC, Noirhomme Q, Fins JJ, Schiff ND, Victor JD (2013) Reanalysis of "Bedside detection of awareness in the vegetative state: a cohort study". Lancet 381(9863):289–291
21. Gibson R, Fernández-Espejo D, Gonzalez-Lara LE, Kwan BY, Lee DH, Owen AM, Cruse D (2014) Multiple tasks and neuroimaging modalities increase the likelihood of detecting covert awareness in patients with disorders of consciousness. Front Hum Neurosci 8:950 http://dx.doi.org/10.3389/fnhum.2014.00950

22. Gabriel D, Henriques J, Comte A, Grigoryeva L, Ortega JP, Cretin E, Brunotte G, Haffen E, Moulin T, Aubry R, Pazart L (2015) Substitute or complement? Defining the relative place of EEG and fMRI in the detection of voluntary brain reactions. Brain Imaging 290:435–44. doi: 10.1016/j.neuroscience.2015.01.053
23. Riganello F, Sannita WG (2009) Residual brain processing in the vegetative state. J Psychophysiol 23(1):18–26
24. Jennett B (2002) The vegetative state. Medical facts, ethical and legal dilemmas. Cambridge University Press, Cambridge, p 223
25. Beaumont JG, Kenealy PM (2005) Incidence and prevalence of the vegetative and minimally conscious states. Neuropsychol Rehabil 15(3–4):184–189
26. van Erp WS, Lavrijsen JC, van de Laar FA, Vos PE, Laureys S, Koopmans RT (2014) The vegetative state/unresponsive wakefulness syndrome: a systematic review of prevalence studies. Eur J Neurol 21:1361–1368
27. Beljaars DE, Valckx WJ, Stepan C, Donis J, Lavrijsen JC (2015) Prevalence differences of patients in vegetative state in the Netherlands and Vienna, Austria: a comparison of values and ethics. J Head Trauma Rehabil 30(3):E57–E60. doi: 10.1097/HTR.0000000000000058
28. Johnson LS (2011) The right to die in the minimally conscious state. J Med Ethics 37(3):175–178
29. Meisel A, Cerminara K (2004) The right to die: the law of end-of-life decision making. Aspen Law and Business, 3rd ed., New York. pp.1374
30. Doerflinger R (2004) Human dignity in the 'vegetative' state. United States Conference of Catholic Bishops, Washington DC
31. Gigli G, Valente M (2006) Quality of life and vegetative state. Proceeding of the XI assembly of the pontificia academia pro vita, Vatican City, pp 147–157
32. Medical aspects of the persistent vegetative state (2). The Multi-Society Task Force on PVS (1994). N Engl J Med 330(22):1572–1579
33. Dolce G, Quintieri M, Serra S, Lagani V, Pignolo L (2008) Clinical signs and early prognosis in vegetative state: a decisional tree, data-mining study. Brain Inj 22(7–8):617–623
34. Liberati G, Hunefeldt T, Olivetti Belardinelli M (2014) Questioning the dichotomy between vegetative state and minimally conscious state: a review of the statistical evidence. Front Hum Neurosci 8:865
35. Sattin D, Covelli V, Pagani M et al (2014) Do diagnostic differences between vegetative state and minimally conscious state patients correspond to differences in functioning and disability profiles? Results from an observational multi-center study on patients with DOC. Eur J Phys Rehabil Med 50(3):309–321
36. Passler M, Riggs RV (2001) Positive outcomes in traumatic brain injury vegetative state: patients treated with bromocriptine. Arch Phys Med Rehabil 82(3):311–315
37. Singh R, McDonald C, Dawson K et al (2008) Zolpidem in a minimally conscious state. Brain Inj 22(1):103–106
38. Ugoya SO, Akinyemi RO (2010) The place of L-dopa/carbidopa in persistent vegetative state. Clin Neuropharmacol 33(6):279–284
39. Whyte J, Katz D, Long D et al (2005) Predictors of outcome in prolonged posttraumatic disorders of consciousness and assessment of medication effects: a multicenter study. Arch Phys Med Rehabil 86(3):453–462
40. Zafonte RD, Watanabe T, Mann NR (1998) Amantadine: a potential treatment for the minimally conscious state. Brain Inj 12(7):617–621
41. Yamamoto T, Katayama Y, Kobayashi K, Oshima H, Fukaya C, Tsubokawa T (2010) Deep brain stimulation for the treatment of vegetative state. Eur J Neurosci 32(7):1145–1151
42. Georgiopoulos M, Katsakiori P, Kefalopoulou Z, Ellul J, Chroni E, Constantoyannis C (2010) Vegetative state and minimally conscious state: a review of the therapeutic interventions. Stereotact Funct Neurosurg 88(4):199–207
43. Ioannidis JP (2005) Why most published research findings are false. PLoS Med 2(8), e124
44. Snyman N, Egan JR, London K et al (2011) Zolpidem for persistent vegetative state–a placebo-controlled trial in pediatrics. Neuropediatrics 41(5):223–227
45. Giacino JT, Whyte J, Bagiella E et al (2012) Placebo-controlled trial of amantadine for severe traumatic brain injury. N Engl J Med 366(9):819–826

46. Crone JS, Schurz M, Holler Y et al (2015) Impaired consciousness is linked to changes in effective connectivity of the posterior cingulate cortex within the default mode network. Neuroimage 110C:101–109
47. Hannawi Y, Lindquist MA, Caffo BS, Sair HI, Stevens RD (2015) Resting brain activity in disorders of consciousness: a systematic review and meta-analysis. Neurology 84(12):1272–1280
48. Goldman S, Schiff ND (2015) Functional imaging to uncover willful brain behavior in non-communicative patients. Neurology 84:1–2
49. Monti MM, Vanhaudenhuyse A, Coleman MR et al (2010) Willful modulation of brain activity in disorders of consciousness. N Engl J Med 362(7):579–589
50. Monti MM, Rosenberg M, Finoia P, Kamau E, Pickard JD, Owen AM (2014) Thalamo-frontal connectivity mediates top-down cognitive functions in disorders of consciousness. Neurology 84(2):167–173
51. Monti MM (2012) Cognition in the vegetative state. Annu Rev Clin Psychol 8:431–454
52. Owen AM, Coleman MR, Boly M, Davis MH, Laureys S, Pickard JD (2006) Detecting awareness in the vegetative state. Science 313:1402
53. Owen AM (2014) Disorders of consciousness: diagnostic accuracy of brain imaging in the vegetative state. Nat Rev Neurol 10(7):370–371
54. Celesia GG, Sannita WG (2013) Can patients in vegetative state experience pain and have conscious awareness? Neurology 80(4):328–329
55. Celesia GG (2013) Conscious awareness in patients in vegetative states: myth or reality? Curr Neurol Neurosci Rep 13(11):395
56. Coleman M, Davis MH, Rodd M, Robson T, Ali A, Owen AM, Pickard JD (2009) Towards the routine use of brain imaging to aid the clinical diagnosis of disorders of consciousness. Brain 132:2541–2555
57. Bruno MA, Majerus S, Boly M, Vanhaudenhuyse A, Schnakers C, Gosseries O, Boveroux P, Kirsch M, Demertzi A, Bernard C et al (2011) Functional neuroanatomy underlying the clinical subcategorization of minimally conscious state patients. J Neurol 259(6):1087–1098
58. Coleman MR, Rodd JM, Davis MH, Johnsrude IS, Menon DK, Pickard JD, Owen AM (2007) Do vegetative patients retain aspects of language comprehension? Evidence from fMRI. Brain 130(Pt 10):2494–2507
59. Di HB, Yu SM, Weng XC, Laureys S, Yu D, Li JQ, Qin PM, Zhu YH, Zhang SZ, Chen YZ (2007) Cerebral response to patient's own name in the vegetative and minimally conscious states. Neurology 68(12):895–899
60. Fernandez-Espejo D, Norton L, Owen AM (2014) The clinical utility of fMRI for identifying covert awareness in the vegetative state: a comparison of sensitivity between 3T and 1.5T. PLoS One 9(4), e95082
61. Owen AM (2015) Using functional magnetic resonance imaging and electroencephalography to detect consciousness after severe brain injury. Handb Clin Neurol 127:277–293
62. Nachev P, Hacker PM (2010) Covert cognition in the persistent vegetative state. Prog Neurobiol 91:68–76
63. Shea N, Bayne T (2010) The vegetative state and the science of consciousness. Br J Philos Sci 61(3):459–484
64. Crone JS, Soddu A, Holler Y et al (2014) Altered network properties of the fronto-parietal network and the thalamus in impaired consciousness. Neuroimage Clin 4:240–248
65. Rodriguez Moreno D, Schiff ND, Giacino J, Kalmar K, Hirsch J (2010) A network approach to assessing cognition in disorders of consciousness. Neurology 75(21):1871–1878
66. Markl A, Yu T, Vogel D, Muller F, Kotchoubey B, Lang S (2013) Brain processing of pain in patients with unresponsive wakefulness syndrome. Brain Behav 3(2):95–103
67. Yu T, Lang S, Vogel D, Markl A, Mueller F, Kotchoubey B (2013) Patients with unresponsive wakefulness syndrome respond to the pain cries of other people. Neurology 80:345–352
68. Kotchoubey B, Merz S, Lang S, Markl A, Muller F, Yu T, Schwarzbauer C (2013) Global functional connectivity reveals highly significant differences between the vegetative and the minimally conscious state. J Neurol 260:975–983

69. Boly M, Faymonville ME, Schnakers C, Peigneux P, Lambermont B, Phillips C, Lancellotti P, Luxen A, Lamy M, Moonen G (2008) Perception of pain in the minimally conscious state with PET activation: an observational study. Lancet Neurol 7:1013–1020
70. Stender J, Kupers R, Rodell A et al (2014) Quantitative rates of brain glucose metabolism distinguish minimally conscious from vegetative state patients. J Cereb Blood Flow Metab 35(1):58–65
71. Staffen W, Kronbichler M, Aichhorn M, Mair A, Ladurner G (2006) Selective brain activity in response to one's own name in the persistent vegetative state. J Neurol Neurosurg Psychiatry 77(12):1383–1384
72. Demertzi A, Gomez F, Crone JS et al (2014) Multiple fMRI system-level baseline connectivity is disrupted in patients with consciousness alterations. Cortex 52:35–46
73. Dehaene S, Changeux J-P (2011) Experimental and theoretical approaches to conscious processing. Neuron 70(2):200–227
74. Sannita WG (2014) Human brain physiology investigated in the disorder of consciousness. Front Neurol 5:211
75. Brooks D (2013) Beyond the brain. New York Times:A21
76. Satel S, Lilienfeld SO (2013) Braihwashed:the seductive appeal of mindless neuroscience. Basic Books, New York, p 288
77. Holland S, Kitzinger C, Kitzinger J (2014) Death, treatment decisions and the permanent vegetative state: evidence from families and experts. Med Health Care Philos 17(3): 413–423
78. Wikler D (1988) Not dead, not dying? Ethical categories and persistent vegetative state. Hastings Cent Rep 18(1):41–47
79. Weir R, Gostin L (1990) Decision to abate life-sustaining treatment for nonautonomous patients. JAMA 264:1846–1853
80. Brody BA (1988) Ethical questions raised by the persistent vegetative patient. Hasting Cent Rep 18:33–37
81. Celesia G (1997) Persistent vegetative state: clinical and ethical issues. Theor Med 18(3): 221–236
82. Cranford R (1988) The persistent vegetative state: the medical reality (getting the facts straight). Hastings Cent Rep 18:27–32
83. Feinberg WM, Ferry PC (1984) A fate worse than death: the persistent vegetative state in childhood. Am J Dis Child 138(2):128–130
84. Williams K (2014) The vegetative state: a fate worse than death. http://www.near-death.com/experiences/suicide10.html
85. Fisher CE, Appelbaum PS (2010) Diagnosing consciousness: neuroimaging, law, and the vegetative state. J Law Med Ethics 38(2):374–385
86. Poldrack RA (2006) Can cognitive processes be inferred from neuroimaging data? Trends Cogn Sci 10(2):59–63
87. Constable C (2010) Withdrawal of artificial nutrition and hydration for patients in a permanent vegetative state: changing tack. Bioethics 26(3):157–163
88. Skene L, Wilkinson D, Kahane G, Savulescu J (2009) Neuroimaging and the withdrawal of life-sustaining treatment from patients in vegetative state. Med Law Rev 17(2):245–261
89. Varelius J (2011) Respect for autonomy, advance directives, and minimally conscious state. Bioethics 25(9):505–515
90. Bernat JL (2009) Ethical issues in the treatment of severe brain injury: the impact of new technologies. Ann N Y Acad Sci 1157:117–130
91. Laureys S, Owen AM, Schiff ND (2004) Brain function in coma, vegetative state, and related disorders. Lancet Neurol 3(9):537–546
92. Reyes-Burke M (2004) The Right to Die vs. The "Life Sentence" for the Minimally Conscious Patient. Public Health Law Policy J 1:47–90
93. Eisenberg J (2008) Schiavo on the cutting edge: functional brain imaging and its impact on surrogate end-of-life decision-making. Neuroethics 1(2):75–83
94. Tolle S (2015) The trouble with advance directives. New York Times. ScienceTimes, D2

95. Kitzinger C, Kitzinger J (2014) Withdrawing artificial nutrition and hydration from minimally conscious and vegetative patients: family perspectives. J Med Ethics 41(2):157–160
96. Giovannetti AM, Leonardi M, Pagani M, Sattin D, Raggi A (2013) Burden of caregivers of patients in vegetative state and minimally conscious state. Acta Neurol Scand 127(1):10–18
97. Leonardi M, Giovannetti AM, Pagani M, Raggi A, Sattin D (2012) Burden and needs of 487 caregivers of patients in vegetative state and in minimally conscious state: results from a national study. Brain Inj 26(10):1201–1210
98. Pagani M, Giovannetti AM, Covelli V, Sattin D, Raggi A, Leonardi M (2013) Physical and mental health, anxiety and depressive symptoms in caregivers of patients in vegetative state and minimally conscious state. Clin Psychol Psychother 21(5):420–426
99. The "death panel" debate (2009) Chicago Tribume. 8/16/2009 ed. Chicago. p 22
100. Paris J (1981) The six-million-dollar woman. Conn Med 45:720–721
101. Kaufman DM, Lipton RB (1992) The persistent vegetative state: an analysis of clinical correlates and costs. N Y State J Med 92(9):381–384
102. Long term care information. U.S. Department of Health and Human Services. http://longtermcare.gov/costs-how-to-pay/costs-of-care/

Chapter 14
Moving Beyond End of Life: The Ethics of Disorders of Consciousness in an Age of Discovery and Uncertainty

L. Syd M. Johnson

Abstract In a time of assured hopelessness, the vegetative state inspired the right to die movement. In the current age of discovery, diagnostic and prognostic uncertainty about disorders of consciousness prevails, while hopelessness is no longer certain. The social, ethical, and legal discourses about these disorders must move beyond end-of-life and right to die issues, to engage with the new uncertainty, and reconsider the status of these patients and the rights that follow from restoring them to humanity and personhood. A less speculative, more empirically informed approach to the ethics of disorders of consciousness—an ethics of uncertainty—is needed.

Keywords Disorders of consciousness • Vegetative state • Unresponsive wakefulness syndrome • Minimally conscious state • Right to die • Neuroethics • Uncertainty

For decades, certainty prevailed concerning disorders of consciousness (DoC). The erosion of that certainty began as a trickle of knowledge that refined and expanded the diagnostic categories. In recent years, a flood of new research and information has rapidly altered the medical and neuroscientific landscape while presenting new and pressing social, ethical, and legal challenges.

14.1 The Age of Hopelessness

In 1972, Jennett and Plum named and described the persistent vegetative state (PVS), a syndrome, previously called *apallic syndrome* or *coma vigile*, marked by chronic unresponsiveness plus eyes-open wakefulness without awareness. Jennett and Plum sought to describe a syndrome that could be diagnosed behaviorally at the bedside and that would not imply any particular etiology or lesion. They fatefully

---

L.S.M. Johnson
Department of Humanities, Michigan Technological University, Houghton, MI 49931, USA
e-mail: lsjohnso@mtu.edu

M.M. Monti, W.G. Sannita (eds.), *Brain Function and Responsiveness in Disorders of Consciousness*, DOI 10.1007/978-3-319-21425-2_14

adopted the term *vegetative* from archaic usage in the *Oxford English Dictionary* to "describe 'an organic body capable of growth and development but devoid of sensation and thought (1764)'" and *vegetate*, meaning "to live a merely physical life, devoid of intellectual activity or social intercourse (1740)" [1]. The term "vegetative" had been used informally to describe comatose patients for a century. Jennett and Plum intended for the term to indicate to doctors a condition in which autonomic functions continue while "the cerebral cortex is not functioning," and they sought terminology that laypersons could understand. Unfortunately, it quickly took on a more pejorative connotation, as patients were referred to as "vegetables," implying that they were not fully human.

Jennett and Plum described PVS patients as hopeless cases, "incapable of communication and without hope of recovery as social human beings." The "hopeful" cases, they said explicitly, are the ones who awaken from coma with signs of cortical functioning, unlike the PVS patients. In this description and designation of the PVS, the idea of *hopelessness* makes an early appearance, and Jennett and Plum were well aware of its ethical and social significance:

> … the immediate issue is to recognize that there is a group of patients who never show evidence of a working mind… Certainly the indefinite survival of patients in this state presents a problem with humanitarian and socioeconomic implications which society as a whole will have to consider. If it were possible to predict soon after the brain damage had been sustained that, in the event of survival, the outcome would be a vegetative, mindless state, then the wisdom of continuing supportive measures could be discussed. [1]

Thus, early in the history of PVS, ethical questions were raised about whether life is worth living for these "mindless" patients, who were viewed as the unfortunate collateral damage of the effort to save the lives of patients with severe brain damage. Saving those who were not hopeless cases meant saving some who were, because outcomes could not be predicted.

That sense of hopelessness continues to be pervasive and persistent in both medical and bioethical discourses. PVS has been called "a fate worse than death" [2], in which maintaining life support is a "futile endeavor" that cannot benefit the patient [3]. PVS patients are frequently characterized as dead, as in this account that describes a doctor as "non-dead" and "lying in an open grave," someone who suffers "the ultimate curse" [4]. It has been argued that the patient *should* be considered dead, because "the body in PVS has lost all forms of personhood" [5]. The humanity of vegetative patients has been challenged on the grounds that "they permanently lack the most essential feature of the human being, which is its undisputed capacity for living the life of a person, or rational self-consciousness" [6]. The perception that the vegetative state is a condition worse than death is widely endorsed by the public [7], with some attributing less *mind* to the vegetative patient than to the dead and viewing them as "more dead than dead" [7]. Many family members view their loved ones as being in an ambiguous state—not straightforwardly alive and not simply dead [7, 8].

A mere 3 years after the Jennett and Plum paper appeared, the Karen Ann Quinlan case emerged as a pivotal legal and bioethical case involving the right to die for PVS patients. In upholding the right of Quinlan's parents to withdraw life support, the New Jersey Supreme Court noted that the state had no important interest in compelling Quinlan to "endure the unendurable, only to vegetate a few measurable months with no realistic possibility of returning to any semblance of cognitive or sapient life" [9]. The court considered her condition to be "debilitated and moribund" and "cruel and most unusual" [9]. Those views were echoed several years later by the United States Supreme Court in the Nancy Cruzan case: "If she has any awareness of her surroundings, her life must be a living hell" [10]. Justice Stevens argued for the propriety of considering quality of life in right to die cases:

> It is appropriate to consider the quality of life in making decisions about the extraordinary medical treatment. Those who have made decisions about such matters without resort to the courts certainly consider the quality of life, and balance this against the unpleasant consequences to the patient. [10]

In the case of Tony Bland, the British High Court concluded that "To his parents and family he is 'dead.' His spirit has left him and all that remains is the shell of his body" [11]. That ruling prompted Jennett to grant the possibility of recovery from the PVS, which he noted would be more curse than blessing: "The recovery of a limited degree of awareness may indeed be worse than non-sentience for the patient." Jennett praised the High Court for judging that it would be in Bland's best interests to have artificial nutrition and hydration withdrawn, noting again that the price of saving those patients with a "reasonable probability of meaningful recovery and of regaining life as a social person" is paid by "patients who die after a fruitless fight in intensive care or survive with irreparable brain damage" [12].

The right to die debate and DoC have thus been entwined from the beginning, and it was the utter hopelessness of the PVS that inspired and gave fuel to those who fought for the legal right to die with dignity, without medical interventions maintaining a life some view to be pointless, perhaps even worse than death. Indeed, when Jennett and Plum wrote of the PVS [1], it *was* a seemingly hopeless condition, an artifact of medical progress that saved many lives, but did so indiscriminately, without the capacity to distinguish those who would go on to have "reasonable" recoveries and those who would remain forever trapped in a meaningless existence.

Twenty-two years after the PVS was named and defined, temporal criteria were established to distinguish the *persistent* vegetative state from the *permanent* vegetative state [13]. By then, it was understood that some patients *do* recover from a VS.[1] Unfortunately, the *persistent* and the *permanent* VS have the same initials, which continues to cause confusion among the public and popular press. The temporally, diagnostically, prognostically, and ethically distinct states continue to be conflated in the bioethics literature as well.

---

[1] The name of the vegetative state has become contentious. Here, VS will refer to the *vegetative state*, without reference to persistence or permanence. PVS will refer only to the *persistent* vegetative state. The term "unresponsive wakefulness syndrome" (UWS) will be discussed below.

14.2 From Certainty to Uncertainty

The diagnostic criteria for the minimally conscious state (MCS) were formally defined in 2002 [14]. The ethical significance of the MCS is difficult to overstate. First, because some patients progress from the VS to the MCS, its existence is proof that the VS is not always permanent, and not so certainly hopeless. Second, because the prospect of further recovery is better in the MCS, it shows that even severe brain injuries are not as hopeless as once thought. Finally, it is evidence of a significant problem of misdiagnosis—one that has resisted improvement for decades. The rate of misdiagnosis in DoC is estimated to be 39–43 % [15, 16], which would be unacceptably high in any area of medicine but is of special concern with DoC, where there is a type II or false-negative error: patients are diagnosed as unconscious when they are in fact conscious. The error occurs because the MCS can be exceedingly difficult to distinguish from the VS and because of a kind of self-fulfilling prophecy: when the VS is believed to be permanent, subtle signs of recovery are ignored or unnoticed. The diagnostic and prognostic error can prompt limitations or withdrawals of care for patients, on the belief that further care would be of no benefit. This creates a cycle of neglect, as patients who are believed to be *permanently* vegetative do not receive the care that might alter their prognosis or uncover signs of recovery.

In recent years, active neuroimaging paradigms using mental imagery tasks have revealed behaviorally unresponsive, covertly aware patients diagnosed in a VS who can willfully modulate their brain activity on command, indicating consciousness [17–19]. Given the problem of misdiagnosis in DoC, the ability to detect a misdiagnosis or effect a rediagnosis through neuroimaging would be tremendously valuable, especially for that subset of patients who regain consciousness undetected. Bedside behavioral diagnosis is susceptible to type II errors but so too neuroimaging. The essential problem, known to philosophers as the problem of other minds, is that the subjective consciousness of others cannot be directly observed. It can only be inferred from indirect evidence, such as behaviors. If neuroimaging provides indirect, objective, and reliable evidence of consciousness in the form of apparently willful *brain behaviors* [20], it does not provide similarly reliable evidence of *un*consciousness. Thus, active neuroimaging paradigms are not immune to false negatives—they might also fail to detect consciousness in some conscious subjects. The "gold standard" for diagnosis of DoC, the bedside behavioral exam, cannot confirm consciousness in behaviorally unresponsive patients, so there is for now no way to completely eliminate false negatives or to disentangle whether the causes of negative neuroimaging findings are a failure of the imaging technique or a confounding factor such as aphasia or agnosia in an individual patient. Nonetheless, from the inference that *some* behaviorally unresponsive patients are conscious and capable of intentional brain behavior, there is a further inference to be made: there must be other patients whose consciousness remains undetected.

Misdiagnosis in patients who are behaviorally unresponsive, or minimally responsive, can take years or decades to discover. It takes on average 2.5 months to diagnose locked-in syndrome, which behaviorally mimics the VS although consciousness is intact; there are cases in the literature of diagnosis after 6 years, during which time the patient was thought to be unconscious [21]. Terry Wallis' progression from VS to MCS remained undetected for almost two decades—until he began to speak—although his family long suspected that he was conscious [22]. There are many cases of unexpected recoveries in the literature, including patients who recover to states of partial disability, functional independence, and return to school or work, even after periods of altered consciousness lasting years [23–25]. Such cases raise questions about diagnostic accuracy, as well as the time-specific permanence of the VS.

In 2010, the European Taskforce proposed changing the name of the VS to unresponsive wakefulness syndrome (UWS), in part to end confusion about the permanent/persistent distinction and in part to ease the stigma attached to the term vegetative, which popularly implies a less-than-human status [26]. The new name would also describe the syndrome's defining features—unresponsiveness and wakefulness—without implying that the patient is unconscious. This is a significant and controversial departure from the Jennett and Plum nomenclature. It grants the inherent uncertainty of unconsciousness, where previously there was implied certainty. UWS also abandons the certainty of permanence in light of the rare cases of late recovery from VS/UWS.

14.3 Certainty and Uncertainty as Ethical Problems

The erosion of diagnostic and prognostic certainty about DoC generates ethical uncertainties. Prognostication is ethically disquieting when it leads to self-fulfilling prophecies that in turn influence *future* prognostications. Physicians tend to overestimate poor outcomes and underestimate good outcomes, particularly when making early predictions [27–29]. Withdrawal of life-sustaining treatment (LST) is the leading factor in neurocritical intensive care unit deaths and precedes up to 70 % of neuro-ICU deaths [30–33]. When a high proportion of deaths follow from decisions to limit or withdraw LST, self-fulfilling prophecies create an epistemic challenge to prognostication: is there high mortality in the population of patients with severe brain injuries, or is high mortality caused by *predictions* of high mortality and morbidity that prompt the withdrawal of care? When predictions of mortality after brain injury are overestimated because of withdrawals of care, a false sense of hopelessness can result.

Given attitudes about the intrinsic badness of VS/UWS, there is a dilemma about what would be the worst possible outcome: life with severe disability or death when life could have been acceptable. This dilemma generates belief in a "window of opportunity," a period of uncertainty soon after acute brain injury, when withdrawal of LST is very likely to lead to death, but when prognosis for survival or recovery with continued care is uncertain [34]. The window of opportunity creates a sense of urgency regarding withdrawal decisions, with the perceived risk being that the patient might recover sufficiently to be independent of LST—thus closing the window—but will survive with severe disability that might not be acceptable to the patient. Waiting for more prognostic certainty risks the closing of the window, and the resulting time pressure is costly, because avoiding the worst outcome for the patient—survival when death would be preferable—requires forgoing any possibility of an acceptable outcome [35]. The short window of opportunity forces early prognostication, which commingles with value judgments concerning acceptable outcomes. Death is a well-defined endpoint, but "poor functional outcomes" are more subjectively evaluative [36]. The prospect of a VS/UWS as one of the "poor functional outcomes"—it is one tick above death on the Glasgow Outcome Scale—certainly factors into decisions to withdraw LST [36]. Pressure to act before the window closes comes not from the possibility of an avoidable death but rather the chance that an *opportunity to die* will be missed [37], resulting in an avoidable survival in a condition *worse than death*. Yet, even if we accept the hopelessness of the *permanent* VS, many decisions to withdraw treatment occur within the first 72 h following injury [32], months before a diagnosis of permanent VS—and hopelessness—can be made. Here the words of Jennett and Plum prove prophetic: if it could be predicted soon after a brain injury that "the outcome would be a vegetative, mindless state, then the wisdom of continuing supportive measures could be discussed" [1]. Given the prevalence of withdrawal of care following severe brain injury, false *hope* appears not to be a problem, but perhaps false *hopelessness* is.

14.4 The Ethics of Uncertainty

Renewed neuroscientific interest in DoC in recent years has greatly enhanced understanding of these disorders, progressing the science from a lengthy age of nihilism and hopelessness to an age of discovery. At the same time, new *uncertainties* about these disorders, and patients, have grown, thus ushering in an age of uncertainty. Corresponding advances in the social, legal, and ethical discourses about DoC have been slow to follow, and likewise a reckoning with the new uncertainty.

Persons with DoC thus remain exceptionally marginalized, set apart from the general population and from other people with brain injuries. Their ontological, diagnostic, and moral segregation has resulted in their contested personhood and moral status, challenges to their right to medical resources [38], and acceptance that the right to die is their only remaining right. DoC reveal how conceptions of personhood or moral status that are contingent on consciousness are deeply vexed, given that unconsciousness is both liable to fluctuation and epistemically uncertain. To qualify loss of personhood as the *permanent* loss of consciousness is also epistemically flimsy and requires admitting that we are wrong 40 % of the time about who is not a person.

The new uncertainties about DoC call for an new, less speculative, more empirically informed ethical approach—an ethics of uncertainty in which the focus is shifted from end of life and the right to die, to other rights that follow from restoring these patients to humanity and personhood: the right to the right diagnosis and empirically informed prognosis (and an acknowledgment of the uncertainty of both), the right to consideration of their actual quality of life (rather than presumed quality of life), and the right to treatments and research that address the needs of patients in states of impaired consciousness. These rights would not erode the right to die. Indeed, providing surrogates with better diagnostic and prognostic information and an informed understanding of quality of life could serve to reinforce decisions and provide the comfort of confidence for families faced with difficult choices.

14.5 From the Right to Die to the Fundamental Right to the Right Diagnosis

Every diagnosis of VS/UWS is not incorrect, nor would the decisions made on behalf of patients always be different with another diagnosis or prognosis. Decisions to limit, withdraw, or continue care are multifactorial and consider the patient's views or values, the needs, beliefs, and values of the patient's family, and the opinions of clinicians. While physicians' estimates of poor outcome are significantly associated with withdrawal of LST early after severe brain injury [29], many families recognize the limitations of predictions for critically ill patients and view uncertainty as an unavoidable reality even as they seek prognostication as part of the decision making process [39]. Nonetheless, there are ethically important concerns—beyond end-of-life decisions—that make it imperative that DoC patients get the right diagnosis and that consciousness, however limited, is recognized. The right to the right diagnosis is fundamental and grounded in the principles of respect for autonomy, beneficence/non-maleficence, and justice.

Principles of beneficence and non-maleficence demand striving for the right diagnosis, because the prognosis for patients in the MCS is considerably better than for the VS/UWS. Discriminating between them can mean the difference between clinical nihilism—doing nothing in the belief that nothing will benefit the patient—and providing just access to care and rehabilitation that might promote further recovery. It is also vital to allow that uncertainty prevails in both diagnosis and prognosis and that some patients who are behaviorally unresponsive may nonetheless be aware and experiencing a life the quality of which remains unknown. There are no first-person accounts of what it's like to be in a VS/UWS, but there are accounts of what it's like to be mistaken for someone in a VS/UWS and what it's like for a patient to be treated like "a vegetable," their personhood disregarded [40–42]. They reveal the devastating physical and emotional toll when the needs of covertly conscious patients are neglected. Investigations into factors like pain perception that affect quality of life have only recently been undertaken in DoC patients. Neuroimaging studies have yielded empirical evidence of pain perception and the potential for suffering, as well as emotional response in the MCS [43–45], demonstrating the need to treat pain or distress or perhaps provide positive and enjoyable stimuli. Such evidence might also inform decisions to end LST if doing so is consistent with the values and desires of patients and their proxies [46].

Persistent uncertainty about quality of life continues to provoke legal and ethical controversy about the withdrawal of care [10, 47]. Quality of life is fundamentally subjective, so certainty about the subjective well-being of another will never be possible. Acknowledging epistemic uncertainty as an ethical constraint does not rule out any options, including withdrawal of LST, but it does force a recognition that *speculation* about quality of life makes a weak foundation for the right to die and cannot undergird indifference to the other rights or needs of DoC patients. Indeed, if we are to respect the autonomy of DoC patients, we do so by providing their surrogates with better diagnostic, prognostic, and quality of life information, so that they can make choices that accord with the values and desires of patients who cannot speak for themselves.

Replacing certainty with uncertainty does not sound like progress, but the current age of uncertainty presents many opportunities, including opportunities to reconsider the dogmas engendered by the false hopelessness and certainties of the past. They were well-meaning certainties, and hopelessness was not a product of bad will but of limited knowledge. The effects—the marginalization, neglect, and dehumanization of DoC patients—can be reversed by attending to both the fact of uncertainty and the ethical obligations that follow from it.

References

1. Jennett B, Plum F (1972) Persistent vegetative state after brain damage: a syndrome in search of a name. Lancet 1:734–736
2. Feinberg WM, Ferry PC (1984) A fate worse than death. The persistent vegetative state in childhood. Am J Dis Child 138(2):128–130
3. Jennett B, Dyer C (1991) Persistent vegetative state and the right to die: the United States and Britain. BMJ 302(6787):1256–1258
4. Crisci C (1995) The ultimate curse. J Med Ethics 21(5):277
5. Ravelingien A, Mortier F, Mortier E, Kerremans I, Braeckman J (2004) Proceeding with clinical trials of animal to human organ transplantation: a way out of the dilemma. J Med Ethics 30(1):92–98
6. Rich BA (1997) Postmodern personhood: a matter of consciousness. Bioethics 11(3&4): 206–216
7. Gray K, Knickman TA, Wegner DM (2011) More dead than dead: perceptions of persons in the persistent vegetative state. Cognition 121:275–280
8. Kuehlmeyer K, Borasio GD, Jox RJ (2012) How family caregivers' medical and moral assumptions influence decision making for patients in the vegetative state: a qualitative interview study. J Med Ethics 38(6):332–337
9. In re Quinlan, 355 A.2d 647, 70 N.J. 10, 70 New Jersey 10 (1976)
10. Cruzan V (1990) Director, Mo. Dept. of Health, 497 U.S. 261, 110 S. Ct. 2841, 111 L. Ed. 2d 224
11. Fergusson A (1992) Letter to the editor. Br Med J 305:1506
12. Jennett B (1992) Letting vegetative patients die: ethical and lawful and brings Britain into line. Br Med J 305:1305–1306

13. Multi-Society Task Force on PVS (1994) Medical aspects of the persistent vegetative state. N Engl J Med 330:1499–1508
14. Giacino JT, Ashwal S, Childs N, Cranford R, Jennett B, Katz DI, Kelly JP et al (2002) The minimally conscious state definition and diagnostic criteria. Neurology 58(3):349–353
15. Schnakers C, Vanhaudenhuyse A, Giacino J, Ventura M, Boly M, Majerus S, Moonen G, Laureys S (2009) Diagnostic accuracy of the vegetative and minimally conscious state: clinical consensus versus standardized neurobehavioral assessment. BMC Neurol 9:35. doi:10.1186/1471-2377-9-35
16. van Erp WS, Lavrijsen JC, Vos PE, Bor H, Laureys S, Koopmans RT (2015) The vegetative state: prevalence, misdiagnosis, and treatment limitations. J Am Med Dir Assoc 16(1): 85e9–85e19
17. Cruse D, Chennu S, Fernández-Espejo D, Payne WL, Young GB, Owen AM (2012) Detecting awareness in the vegetative state: electroencephalographic evidence for attempted movements to command. PLoS One 7(11):e49933
18. Monti MM, Vanhaudenhuyse A, Coleman MR, Boly M, Pickard JD, Tshibanda L, Owen AM, Laureys S (2010) Willful modulation of brain activity in disorders of consciousness. N Engl J Med 362(7):579–589
19. Owen AM, Coleman MR, Boly M, Davis MH, Laureys S, Pickard JD (2006) Detecting awareness in the vegetative state. Science 313:1402
20. Monti MM, Coleman MR, Owen AM (2010) Behavior in the brain: using functional neuroimaging to assess residual cognition and awareness after severe brain injury. J Psychophysiol 24(2):76–82
21. Laureys S, Pellas F, Van Eeckhout P, Ghorbel S, Schnakers C, Perrin F, Berré J, Faymonville M, Pantke K, Damas F, Lamy M, Moonen G, Goldman S (2005) The locked-in syndrome: what it is like to be conscious but paralyzed and voiceless? In: Laureys S (ed) Prog Brain Res 150(34):495–511
22. Wijdicks EFM (2006) Minimally conscious state vs persistent vegetative state: the case of Terry (Wallis) vs the case of Terri (Schiavo). Mayo Clin Proc 81(9):1155–1158
23. Edlow BL, Giacino JT, Hirschberg RE, Gerrard J, Wu O, Hochberg LR (2013) Unexpected recovery of function after severe traumatic brain injury: the limits of early neuroimaging-based outcome prediction. Neurocrit Care 19(3):364–375
24. Katz DI, Polyak M, Coughlan D, Nichols M, Roche A (2009) Natural history of recovery from brain injury after prolonged disorders of consciousness: outcome of patients admitted to inpatient rehabilitation with 1–4 year follow-up. Prog Brain Res 177:73–88
25. Steppacher I, Kaps M, Kissler J (2014) Will time heal? A long-term follow-up of severe disorders of consciousness. Ann Clin Transl Neurol 1(6):401–408
26. Laureys S, Celesia GG, Cohadon F, Lavrijsen J, León-Carrión J, Sannita WG, Sazbon L et al (2010) Unresponsive wakefulness syndrome: a new name for the vegetative state or apallic syndrome. BMC Med 8(1):68
27. Kaufmann MA, Buchmann B, Scheidegger D, Gratzl O, Radü EW (1992) Severe head injury: should expected outcome influence resuscitation and first-day decisions? Resuscitation 23(3):199–206
28. Rabinstein AA, Claude Hemphill J (2010) Prognosticating after severe acute brain disease Science, art, and biases. Neurology 74(14):1086–1087
29. Rocker G, Cook D, Sjokvist P, Weaver B, Finfer S, McDonald E, Marshall J et al (2004) Clinician predictions of intensive care unit mortality. Crit Care Med 32(5):1149–1154
30. Côte N, Turgeon AF, Lauzier F, Moore L, Scales DC, Bernard F, Zarychanski R et al (2013) Factors associated with the withdrawal of life-sustaining therapies in patients with severe traumatic brain injury: a multicenter cohort study. Neurocrit Care 18(1):154–160
31. Izzy S, Compton R, Carandang R, Hall W, Muehlschlegel S (2013) Self-fulfilling prophecies through withdrawal of care: do they exist in traumatic brain injury, too? Neurocrit Care 19(3):347–363
32. Turgeon AF, Lauzier F, Simard JF, Scales DC, Burns KE, Zygun DA et al (2011) Mortality associated with withdrawal of life-sustaining therapy for patients with severe traumatic brain injury: a Canadian multicentre cohort study. Can Med Assoc J 183(14):1581–1588

33. Verkade MA, Epker JL, Nieuwenhoff MD, Bakker J, Kompanje EJ (2012) Withdrawal of life-sustaining treatment in a mixed intensive care unit: most common in patients with catastrophic brain injury. Neurocrit Care 16(1):130–135
34. Wilkinson D (2011) The window of opportunity for treatment withdrawal. Arch Pediatr Adolesc Med 165(3):211–215
35. Kitzinger J, Kitzinger C (2013) The 'window of opportunity' for death after severe brain injury: family experiences. Sociol Health Illn 35(7):1095–1112
36. Geurts M, Macleod MR, van Thiel GJ, van Gijn J, Kappelle LJ, van der Worp HB (2014) End-of-life decisions in patients with severe acute brain injury. Lancet Neurol 13(5):515–524
37. Cochrane TI (2009) Unnecessary time pressure in refusal of life-sustaining therapies: fear of missing the opportunity to die. Am J Bioeth 9(4):47–54
38. Johnson LSM (2010) Withholding care from vegetative patients: the social and financial costs. Bioeth Forum. http://www.thehastingscenter.org/Bioethicsforum/Post.aspx?id=4789&blogid=140&terms=withholding+care+and+%23filename+*.html. Accessed 16 Mar 2015
39. Zier LS, Burack JH, Micco G, Chipman AK, Frank JA, Luce JM, White DB (2008) Doubt and belief in physicians' ability to prognosticate during critical illness: the perspective of surrogate decision makers. Crit Care Med 36(8):2341–2347
40. Macniven JA, Poz R, Bainbridge K, Gracey F, Wilson BA (2003) Emotional adjustment following cognitive recovery from 'persistent vegetative state': psychological and personal perspectives. Brain Inj 17(6):525–533
41. Tavalaro J, Tayson R (1997) Look up for yes. Kodansha International, New York
42. Wilson BA, Gracey F (2001) Cognitive recovery from 'persistent vegetative state': psychological and personal perspectives. Brain Inj 15(12):1083–1092
43. Bekinschtein T, Niklison J, Sigman L, Manes FRLJA, Leiguarda R, Armony J, Owen A, Carpintiero S, Olmos L (2004) Emotion processing in the minimally conscious state. J Neurol Neurosurg Psychiatry 75(5):788
44. Schnakers C, Chatelle C, Demertzi A, Majerus S, Laureys S (2012) What about pain in disorders of consciousness? AAPS J 14(3):437–444
45. Schnakers C, Chatelle C, Vanhaudenhuyse A, Majerus S, Ledoux D, Boly M, Bruno MA et al (2010) The Nociception Coma Scale: a new tool to assess nociception in disorders of consciousness. Pain 148(2):215–219
46. Johnson LSM (2013) Can they suffer? The ethical priority of quality of life research in disorders of consciousness. Bioeth Forum 6(4):129–136
47. Glannon W (2013) Burdens of ANH outweigh benefits in the minimally conscious state. J Med Ethics. doi:10.1136/medethics-2012-100882

Chapter 15
Mesocircuit Mechanisms Underlying Recovery of Consciousness Following Severe Brain Injuries: Model and Predictions

Nicholas D. Schiff

Abstract This chapter reviews the mesocircuit hypothesis for the recovery of consciousness after severe brain injuries. Building up from a single premise that multifocal brain injuries typically produce widespread neuronal death and dysfunction, the mesocircuit model proposes a lawful circuit-level mechanism on the basis of known anatomical and physiological specializations of the corticothalamic system. The hypothesis is deeply generative of testable predictions at the mesocircuit and cellular level across multiple cerebral structures. Recent tests of some of these predictions are discussed as well as next steps in evaluating theoretical implications of the mesocircuit model.

Keywords Mesocircuit • Central thalamus • Arousal systems • EEG • Deafferentation • Brain injury

15.1 The Mesocircuit Hypothesis for Recovery of Consciousness After Severe Brain Injuries

Severe brain injuries impair consciousness in many different contexts in which bi-hemispheric cerebral function may be altered by enduring loss or dysfunction of many neuronal populations (see Posner et al. [1] for a comprehensive review). Recovery of consciousness following such brain injuries has been proposed to result from restoration of activity across a defined mesocircuit within the anterior forebrain that is particularly vulnerable to the effects of multifocal structural or functional deafferentation [2–4]. Unstable levels of behavioral responsiveness associated with initial recovery of consciousness after coma in the context of behavioral signs of consciousness emerging in the minimally conscious state (MCS) and in patients recovering just past the

N.D. Schiff, MD
The Jerold B. Katz Professor of Neurology and Neuroscience, Feil Family Brain and Mind Research Institute, Weill Cornell Medical College, New York, NY, 10065, USA
e-mail: nds2001@med.cornell.edu

M.M. Monti, W.G. Sannita (eds.), *Brain Function and Responsiveness in Disorders of Consciousness*, DOI 10.1007/978-3-319-21425-2_15

border of MCS in confusional state can be understood in the context of dynamical shifts in patterns of activity across this anterior forebrain mesocircuit [5].

In brief, the mesocircuit hypothesis (Fig. 15.1) starts with the basic observation that all severe brain injuries share common pathological substrates of widespread functional or structural deafferentation of neocortical, striatal, and thalamic neurons that maximally impacts survival or function of central thalamic neurons [6–8]. The primary consequence of broad deafferentation following diffuse multifocal patterns of corticothalamic disconnection is disfacilitation, a passive hyperpolarization of neurons that reduces their firing rates; experimental studies demonstrate that powerful consequences on background firing rates occur even with disfacilitation producing only modest reductions measured in cerebral blood flow [9]. A marked reduction of global brain synaptic background activity results from the loss of the primarily long-range connections across the forebrain which are dominantly excitatory projections to neocortical, striatal, and thalamic neurons (Fig. 15.1A). The overall impact of widespread disfacilitation is a sharp reduction in cerebral metabolic rate which is considered the common denominator in all disturbances of consciousness following severe brain injuries. The broad reduction of neuronal activity across the mesocircuit is predicted to be specifically graded by a result of two core interacting effects. Most important and general is the role of deafferentation of the neurons of the central thalamus. These central thalamic neurons are unique in providing, in the aggregate, the shortest point-to-point connections across the entire corticothalamic system and as consequence suffer the greatest impact of multifocal deafferentation [8, 10]. Moreover, their functional role in the wakeful state is in adjusting overall levels of activity in the frontostriatal systems to respond to challenges of increasing cognitive load, stress, or fatigue (see Schiff [10] for comprehensive review). Thus, reduction of the output of these neurons is anticipated to have a negative synergistic effect across the corticothalamic system, as powerful modulatory excitatory effects across the cortex and striatum are withdrawn. A second interacting mechanism is the anticipated complete shutdown of firing of the medium spiny neurons (MSNs) of the striatum as a result of their high-threshold UP state and dependence on high-frequency inputs from corticostriatal and thalamostriatal afferents to maintain their firing rates [11]. Shutdown of the striatal MSNs produces disinhibition of the globus pallidus interna and would add subsequent active inhibition of the already disfacilitated central thalamus and possibly a synergistic effect of inhibition of outflow of pedunculopontine nucleus input to the central thalamus further hyperpolarizing these thalamic neurons [3, 12].

There is further a direct relationship between graded reduction of activity across the anterior forebrain mesocircuit and the existing and compelling evidence for a graded relationship of frontoparietal activity in outcomes of severe brain injuries, particularly resting activity within the posterior medial complex of the parietal cortex [13]. Fernandez-Espejo et al. [14] reported that diagnostic categories of disorders of consciousness can be stratified by diffusion tensor imaging (DTI) of connections between the thalamus and the posterior medial complex (PMC). Anatomical and physiological studies show that the central thalamus and the posterior medial cortical regions have strong functional connections. Track-tracing studies in nonhuman primates demonstrate that the anterior intralaminar nuclei provide a substrate through

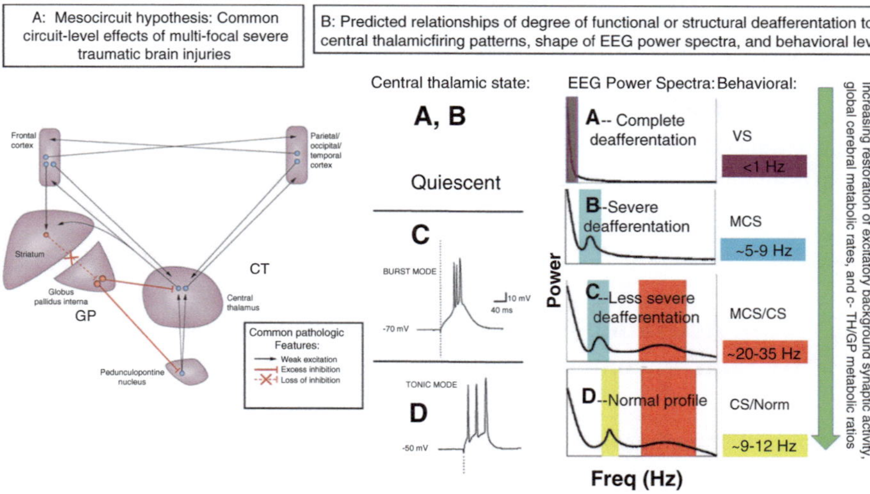

Fig. 15.1 (Panel **a**) The mesocircuit hypothesis [2–5, 15, 16] proposes that, based on known pathologic changes after trauma, hypoxia, or multifocal ischemia, reduction of thalamocortical and thalamostriatal outflow following deafferentation and loss of neurons in central thalamus withdraws important afferent drive to frontal cortex and the medium spiny neurons (MSNs) of the striatum (Adapted from Giacino et al. [12]). Impact across the frontal cortical systems is maximal because of the density of central thalamic projections and a contribution from a failure of MSNs to reach firing threshold because of their requirement for high levels of synaptic background activity [11]. (Panel **b**) Relationship of degree of deafferentation in disorders of consciousness after severe brain injuries, central thalamic state, and electroencephalogram patterns [17]. At varying levels of functional or structural deafferentation, changes in central thalamic firing patterns are expected to lawfully appear and associate with neocortical dynamics measured in the EEG: quiescence in the setting of very severe structural or functional deafferentation (*A*, *B*); bursting (*C*) in setting of significant deafferentation [18], and tonic firing mode with recovery (*D*). Expected changes in EEG spectra are linked to increases in overall level of excitatory drive to neocortical neurons with increasing central thalamic activity and overall shifts of cerebral metabolism that are predicted to arise in a sequence in recovery and to grade with levels of structural or functional deafferentation and behavioral outcome (*green arrow* running vertically down). At the extreme of complete functional deafferentation of neocortical neurons, very low average membrane potentials (AMPs) of neocortical neurons of ~−70 mV are measured and a concentration of spectral power <1 Hz is expected [19] as shown in the "A" spectrum. Patients with very severe cerebral deafferentation who remain in vegetative state may show this pattern (see, e.g., Fig. 15.2a). With sufficient afferent input to place the AMP in the range ~−60–65 mV, spontaneous oscillations of Layer V pyramidal cells [20] produce ~5–9 Hz oscillations. This mechanism is proposed to account for a ~7 Hz oscillation that dominates the resting activity of the EEG in the "B" spectrum as observed in some MCS patients [5]. In both the *A* and *B* setting, the central thalamus is predicted to be quiescent (direct measurements of central regions of 6 thalami in 3 MCS patients from our database support this expectation, Giacino et al. [21, 22]). In the setting of more normal levels of global background synaptic activity and cerebral metabolism, more deafferented regions of the thalamus are expected to burst and produce co-existing ~5–9 and ~15–40 Hz EEG oscillations local to cortical projection regions [18, 23], seen in the "C"-type spectrum. Normal EEG background activity ("D") is by contrast, globally dominated by 8–12 Hz and features within the ~15–40 Hz range [24]. The mesocircuit hypothesis predicts that more *functionally deafferented* neurons will be concentrated in frontocentral regions; this specific spatial localization prediction has been verified for a set of zolpidem responsive subjects with "B"-type spectra [5] and traumatic brain injury patients with spontaneous recovery of communication and higher cognition [25]

their broad innervations of all components of the posterior medial complex (PMC) of the parietal lobe [26]. The human central lateral nucleus and surrounding paralaminar regions receive the heaviest innervations of both brainstem and basal forebrain cholinergic systems (Heckers et al. [27]), and studies of pharmacological stimulation during human anesthesia demonstrate that brief recovery of consciousness is associated with selective significant increases regional blood flow to thalamus and precuneus compared with wakeful baseline when administering a systemic cholinergic agonist [28].

15.2 Bulk Activation of Anterior Forebrain Mesocircuit by Pharmacological and Other Interventions

The impact of several pharmacologic interventions and central thalamic deep brain stimulation on some patients with severe brain injuries can be explained as a result of bulk activation of the anterior forebrain mesocircuit [29]. A variety of specific pharmacological interventions are known to be effective in improving behavioral responsiveness in some severely brain-injured patients, including the dopaminergic agonists such as bromocriptine, apomorphine, L-Dopa, and amantadine [29, 30]. Dopaminergic tone is the other dominant modulator of the output of the medium spiny neurons, and the mesocircuit model would predict activation of the anterior forebrain in cases where sufficient afferent input remained capable of turning on the MSNs once dopamine levels were restored within a deafferented striatum. Direct modulation of mesial frontal cortical neurons by dopaminergic agonists alone would likely be insufficient as an explanation as neuromodulators in general do not drive firing rates. Release of the glutamatergic afferents of the central thalamus, however, by release of the active pallidal inhibition is predicted to restore anterior forebrain activity across the loop connections of the frontal cortex, striatum, pallidum, and central thalamus.

Amantadine has been shown to be a consistently effective medication in improving behavioral function in disorders of consciousness [22], and its strong and broad effects likely reflect a variety of actions [29]. Perhaps most important is its role as an NMDA antagonist, like ketamine which produces broad cortical excitation [2]. Amantadine may produce bulk activation of the mesocircuit via a combination of NMDA antagonism and dopaminergic as well as noradrenergic agonism [29]. Schnakers et al. [30] demonstrated that effective amantadine response in a minimally conscious state patient reliably associated with broad medial and lateral frontal cortical increase in cerebral metabolism. The paradoxical phenomenon of behavioral facilitation observed in some severely brain patients with administration of the sedative agent zolpidem (a GABA-A alpha 1 positive allosteric modulator) may reflect a similar multiplicity of sites of action across the mesocircuit including a critical effect of suppressing active pallidal inhibition [4], direct modulation of cortical and striatal populations, and even a possible role for ephaptic excitation of axons (see Williams et al. [5] for detailed discussion of these mechanisms and their relationship mesocircuit components).

Bulk activation of the anterior forebrain mesocircuit can also be achieved via direct activation of the glutamatergic afferents of the central thalamus with consequent behavioral improvements in severe brain injury [31]. In a single-subject study activation of midline central regions measured by evoked responses obtained from DBS electrodes showed that effective contacts generated bilateral patterns of activation of the electroencephalogram over frontocentral and frontotemporal electrodes, EEG [31]. Of note, evoked cerebral activity from the human central thalamus also showed localized activation of electrodes over the medial parietal cortical regions in this subject supporting the strong functional connectivity between the anterior forebrain mesocircuit and the medial parietal cortex through the central thalamus discussed above (see Laureys and Schiff [13] for further discussion).

15.3 Testing Specific Model Predictions

Recent studies have tested and confirmed several predictions of the mesocircuit hypothesis related to structural injury patterns, cerebral metabolic profiles, and patterns of resting electrophysiological activity in the severely injured brain. In a series of studies, Lutkenhoff et al. [32, 33] elegantly demonstrated that tissue atrophy within the basal ganglia and thalamus systematically associated with clinical measures of awareness and wakefulness providing evidence that the anatomical integrity of these key subcortical components of the anterior forebrain mesocircuit alone is a sensitive marker for behavioral outcome. Physiological studies have also addressed more specific predictions of the mesocircuit hypothesis. Drover et al. proposed an "ABCD" classification scheme that organizes the relationship of the degree of deafferentation in disorders of consciousness after severe brain injuries, central thalamic state, and expected electroencephalogram patterns as shown in Fig. 15.1B [17]. The mesocircuit hypothesis makes several predictions that can be precisely framed using this ABCD model.

The ABCD model is a coarse graded linear organization of degree of cerebral deafferentation and expected effect on neocortical neurons measured in the EEG. One key prediction is that graded functional deafferentation across the neocortex should consistently associate with a anterior-posterior gradient of greater to lesser functional deafferentation producing a lawful progression of patterns of neuronal firing rates reflected in the shape of the EEG power spectrum that strictly covary with regional and global patterns of cortical metabolic activity from most severely downregulated to fully recovered (Fig. 15.1). Beginning with the extreme case of nearly complete cortical deafferentation, the conditions of the experimental slab preparation [19] are approximated where neocortical neurons remain with an average membrane potential of ~−70 mV and population activity reflected in the EEG power spectrum is restricted to frequencies of ~1 Hz. Patients with very severe anoxic brain injuries in vegetative state may demonstrate these "A-type" features of EEG power spectra (See Fig. 15.2A and Schiff et al. [16]). In the healthy intact cerebral cortex without deafferentation, the resting average membrane potential is

~−55 mV [34], and population activity reflected in EEG power spectra show a peak in the alpha (8–12 Hz) along with variable number of peaks in higher frequency ranges ("D-type" spectra, see Fig. 15.2B).

In between the end points of "A"- and "D"-type spectra, two distinct grid points in the model attach to measured physiological phenomena at the neocortical cellular and population level across the corticothalamic system. Since it is reasonable to assume that with partial but nonetheless severe deafferentation, the average membrane potential in neocortical neurons is intermediate between the extremes of ~−70 and ~−55 mV; it is useful to consider specific physiological regimes associated with such conditions. Silva et al. [20] found that at the intermediate level of −60 to −65 mV, a specific class of neocortical neurons in Layer V will spontaneously burst for seconds at a rate of 7.4 +/− 0.6 Hz as a result of intrinsic membrane properties. In a case series of three zolpidem responsive patients with varying etiologies of severe structural deafferentation [5], all subjects demonstrated peaks in the power spectrum at ~7.4 Hz that: (1) appeared most prominently during washout of the drug and with longer times off of the medication, (2) showed a very strong anterior-posterior gradient with small standard deviations over frontal cortical areas, (3) demonstrated sharp increases in inter- and intra-hemispheric coherence, and (4) showed reliable transitions for "B"- to "D"-type spectra associated with behavioral

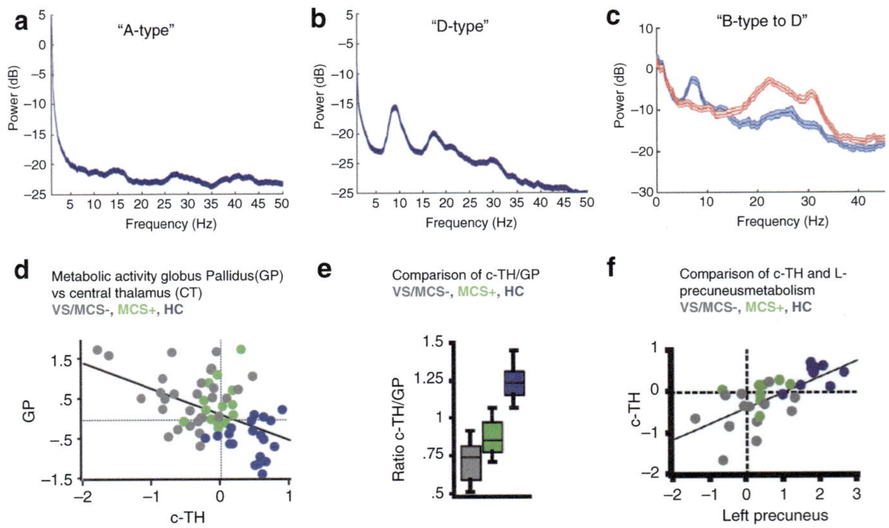

Fig. 15.2 (**a**) Example of an "A"-type spectrum from patient in post-anoxic vegetative state (cf. Fig. 15.1). (**b**) Example of healthy wakeful EEG spectrum ("D-Type"). (**c**) Example of a transition from "B"- to "D"-type spectrum in a zolpidem responsive subject [5], blue spectrum reflects off zolpidem baseline ("B"), red spectrum shows first hour on zolpidem ("D"). (**d**, **e**) Verification of a reversal of metabolic activity ratio of globus pallidus and central thalamus in severe brain injury and grading with behavioral outcome [15]. (**f**) Correlation of ratio of central thalamic and left precuneus metabolism with behavioral outcome following severe brain injury [29] (Figure elements adapted from Schiff et al. [16] (**a**, **b**); Fridman et al. [15] (**d**, **e**) and Fridman and Schiff [29] (**f**))

recovery (see, e.g., Fig. 15.2C). Collectively, these findings support the existence of a "B"-type regime in which an isolated peak within the ~3–7 Hz range of the power spectrum of a very deafferented cortex reflects passively activated intrinsic firing from Layer V neurons (which are the cortical output layer cells) that produces broad synchronization across the hemispheres likely via a weak diffuse coupling ("Huygens clock") mechanism [5]. In the "B"-type regime, central thalamic neurons are proposed to be quiescent, accounting for the dominance of passive driving of neocortical activity by the Layer V cell bursting. In support of this prediction, in one subject studied with positron emission tomography, the off zolpidem state showed a 50 % reduction in cerebral metabolic activity [5]. In addition, separate studies of three human subjects in minimally conscious state that allowed correlation of similar metabolic profiles with direct sampling of extracellular recordings in the thalami showed a dominance of thalamic quiescence [21].

The other intermediate regime between "B" and "D" reflects the partial recovery of normal wakeful tonic activity in the central thalamus (Fig. 15.1B). This "C"-type regime has been proposed by Llinas and colleagues [18, 35] as a "thalamocortical dysrhythmia" in which central thalamic neurons produce low-frequency bursting patterns (restricted to sleep states in the healthy brain) during wakefulness and generate both ~3–7 Hz and higher frequency features in the EEG from overlying cortex due to effects of lateral inhibition. The "C"-type pattern is thus predicted to appear in setting of more preserved cerebral metabolism and to be generated over cortical projection regions of partially deafferented thalamus. Drover et al. [36] have reported evidence for such "C"-type regimes in patients with severe brain injuries.

Several studies of cerebral metabolism or blood flow show evidence for the specific regional changes in cerebral metabolism and blood flow predicted by the mesocircuit model [15, 37, 38]. A key prediction of the mesocircuit model is the expectation of a reversal of the relative metabolic pattern of mean uptake values (mn-UV) from the globus pallidus (GP) and central thalamus (c-TH) from the patient subjects compared with healthy controls; in a study of severely brain-injured subjects from a mixed population of etiology of injuries, this relationship was verified using positron emission tomography techniques [15]. Figure 15.2D shows group data displaying mean uptake values of glucose metabolism within the central thalamus and globus pallidus from ten normal subjects (Blue) and 24 patients (Gray, Green) with severe brain injuries (see Fridman et al. [15] for details). The ratio of central thalamus and globus pallidus metabolism demonstrated a graded relationship with outcome (Fig. 15.2E), suggesting its possible use as a measure to track individual subjects' recovery. In the context of the overall theoretical framework, a lawful covariance of relationships of this ratio, spectral measures, and behavioral level would be expected (Fig. 15.1B). In addition, Fig. 15.2F shows that in the same subjects, central thalamic metabolism correlated with left precuneus metabolism and behavioral level supporting the linkage of the anterior forebrain mesocircuit with this key node in the frontoparietal network known to correlate with outcomes in disorders of consciousness [13]. Similar studies of cerebral blood flow utilizing arterial spin labeling techniques [38] also demonstrated that global cerebral blood flow can distinguish patients in MCS and CS from healthy volunteers and suggest a relatively greater anterior forebrain downregulation of blood flow. Finally, a separate case series of three zolpidem responsive minimally conscious state (MCS) subjects, who all showed from MCS with drug administration showed increased cerebral metabolism within the medial and lateral frontal cortices as the group result in response to zolpidem [37].

15.4 Next Steps

The most immediate next step for testing of predictions of the mesocircuit hypothesis is aimed at verifying and quantifying expected spatiotemporal patterns of dynamics that should be differentially associated with the ABCD categories. Nontrivial predictions exist for the "B"-, "C"-, "D"-type spectra in terms of the global EEG dynamics associated with each type and the physiological conditions expected to generate them. To test these predictions, multivariate EEG analysis methods are required. A well-suited technique is the hierarchical decomposition analysis (HD) method [39] which reorganizes multivariate EEG time series data according to a global model of causal structure. In contrast to principal components analysis (PCA), which parses the EEG into components in terms of variance explained, HD reorganizes the EEG into a hierarchy of components in which the higher components drive the lower ones, but not vice versa. Variance explained does not play a role in HD. In preliminary studies using HD analysis of EEG, we have found evidence in support of some of these predictions [17, 25, 36, 40]. Conte et al. [40] identified a consistent pattern of organization of 8–12 Hz activity in the first hierarchical component of healthy volunteers resting wakeful EEG signals. In preliminary studies of patient subject data, Drover et al. [36] found evidence supporting the anticipated global dynamics of both "B"- and "C"-type spectra. Specifically, for the "B" condition, the dynamical homogeneity of the process and the expected low level of brain activity (i.e., diffusely coupled ~7.4 Hz activity propagating over a quiet cerebral cortex) anticipate that the first hierarchical component is dominated by a spectrum with a ~7.4 Hz peak that is generated by a wide expanse of cortical tissue. For the "C" condition, the "C-type" spectrum is anticipated to be locally generated and dynamically isolated. Examples in support of both of these predictions have been reported [36].

More direct assessments and rigorous test of these model predictions are also possible. Additional studies when possible can correlate measurement of thalamic neuronal activity patterns and EEG spectra (cf. [21, 31]). In vivo animal studies could further test the correlation of extracellular thalamic single-unit and population activity with concomitant intracellular recordings from neocortical neurons (cf. methods in Timofeev et al. [19]) to precisely evaluate the lawfulness of covariation of central thalamic regimes, neocortical average membrane potentials, and local population activity reflected in power spectrum of the cortical local field potential. Finally, the mesocircuit hypothesis frames a template for the examination of possible subcellular mechanisms of neuronal functional impairments altering circuit dynamics (e.g., inflammation, energetics, and metabolic factors).

References

1. Posner J, Saper C, Schiff N, Plum F (2007) Plum and Posner's stupor and coma. Oxford University Press, New York
2. Brown EN, Lydic R, Schiff ND (2010) General anesthesia, sleep, and coma. N Engl J Med 363:2638–2650
3. Schiff ND (2010) Recovery of consciousness after brain injury: a mesocircuit hypothesis. Trends Neurosci 33:1–9
4. Schiff ND, Posner JB (2007) Another "awakenings". Ann Neurol 62:5–7
5. Williams ST et al (2013) Common resting brain dynamics indicate a possible mechanism underlying zolpidem response in severely brain-injured subjects. Elife 2:e01157
6. Adams JH, Graham DI, Jennett B (2000) The neuropathology of the vegetative state after acute insult. Brain 123:1327–1338
7. Jennett B et al (2001) Neuropathology in vegetative and severely disabled patients after head injury. Neurology 56:486–490
8. Maxwell WL et al (2006) Thalamic nuclei after human blunt head injury. J Neuropathol Exp Neurol 65(5):478–488
9. Gold L, Lauritzen M (2002) Neuronal deactivation explains decreased cerebellar blood flow in response to focal cerebral ischemia or suppressed neocortical function. Proc Natl Acad Sci 99:7699–7704
10. Schiff ND (2008) Central thalamic contributions to arousal regulation and neurological disorders of consciousness. Ann N Y Acad Sci 1129:105–118
11. Grillner S, Hellgren J, Ménard A, Saitoh K, Wikström MA (2005) Mechanisms for selection of basic motor programs—roles for the striatum and pallidum.Trends Neurosci 28:364–370
12. Giacino JT, Fins JJ, Laureys S, Schiff ND (2014) Disorders of consciousness after acquired brain injury: the state of the science. Nat Rev Neurol 10(2):99–114
13. Laureys S, Schiff ND (2012) Coma and consciousness: paradigms (re)framed by neuroimaging. Neuroimage 61:478–491
14. Fernandez-Espejo D, Soddu A, Cruse D, Palacios EM et al (2012) Role for the default mode network in the bases of disorders of consciousness. Ann Neurol 72:335–343
15. Fridman EA, Beattie BJ, Broft A, Laureys S, Schiff ND (2014) Regional cerebral metabolic patterns demonstrate the role of anterior forebrain mesocircuit dysfunction in the severely injured brain. Proc Natl Acad Sci U S A 111(17):6473–6478
16. Schiff ND, Nauvel TN, Victor JD (2014) Large-scale brain dynamics in disorders of consciousness. Curr Opin Neurobiol 25:7–14
17. Drover JD et al (2011) Are low frequency oscillations in the EEG of severely injured brains a marker for functional reserve of cortical neurons? Society for neuroscience 39th annual meeting. Abstract 659.8
18. Llinás RR, Ribary U, Jeanmonod D, Kronberg E, Mitra PP (1999) Thalamocortical dysrhythmia: a neurological and neuropsychiatric syndrome characterized by magnetoencephalography. Proc Natl Acad Sci 96:15222–15227
19. Timofeev I, Grenier F, Bazhenov M, Sejnowski TJ, Steriade M (2000) Origin of slow cortical oscillations in deafferented cortical slabs. Cereb Cortex 10:1185–1199
20. Silva LR, Amitai Y, Connors BW (1991) Intrinsic oscillations of neocortex generated by layer 5 pyramidal neurons. Science 251:432–435
21. Giacino J, Fins JJ, Machado A, Schiff ND (2012) Central thalamic deep brain stimulation to promote recovery from chronic posttraumatic minimally conscious state: challenges and opportunities. Neuromodulation 15(4):339–349

22. Giacino JT et al (2012) Placebo-controlled trial of amantadine for severe traumatic brain injury. N Engl J Med 366:819–826
23. Llinás R, Urbano FJ, Leznik E, Ramírez RR, van Marle HJF (2005) Rhythmic and dysrhythmic thalamocortical dynamics: GABA systems and the edge effect. Trends Neurosci 28:325–333
24. Schomer DL, Da Silva FL (2012) Niedermeyer's electroencephalography: basic principles, clinical applications, and related fields. Wolters Kluwer Health, Philadelphia
25. Conte MM et al (2010) Longitudinal changes in the EEG spectrum during recovery after severe brain injury. Society for neuroscience 39th annual meeting. Abstract 659.8
26. Buckwalter JA, Parvizi J, Morecraft RJ, van Hoesen GW (2008) Thalamic projections to the posteromedial cortex in the macaque. J Comp Neurol 507:1709–1733
27. Heckers S, Geula C, Mesulam MM (1992) Cholinergic innervation of the human thalamus: dual origin and differential nuclear distribution. J Comp Neurol 325:68–82
28. Xie G, Deschamps A, Backman SB et al (2011) Critical involvement of the thalamus and precuneus during restoration of consciousness with physostigmine in humans during propofol anaesthesia: a positron emission tomography study. Br J Anaesth 106:548–557
29. Fridman EA, Schiff ND (2014) Neuromodulation of the conscious state following severe brain injuries. Curr Opin Neurobiol 29C:172–177
30. Schnakers C et al (2008) Measuring the effect of amantadine in chronic anoxic minimally conscious state. J Neurol Neurosurg Psychiatry 79(2):225–227
31. Schiff ND et al (2007) Behavioural improvements with thalamic stimulation after severe traumatic brain injury. Nature 448:600–603
32. Lutkenhoff ES et al (2015) Thalamic and extrathalamic mechanisms of consciousness after severe brain injury. Ann Neurol 78(1):68–76
33. Lutkenhoff ES, McArthur DL, Hua X, Thompson PM, Vespa PM, Monti MM (2013) Thalamic atrophy in antero-medial and dorsal nuclei correlates with six-month outcome after severe brain injury. Neuroimage Clin 3:396–404
34. Steriade M, Timofeev I, Grenier F (2001) Natural waking and sleep states: a view from inside neocortical neurons. J Neurophysiol 85:1969–1985
35. Llinás R, Urbano FJ, Leznik E, Ramírez RR, van Marle HJ (2005) Rhythmic and dysrhythmic thalamocortical dynamics: GABA systems and the edge effect. Trends Neurosci 28(6): 325–333
36. Drover JD, Conte MM, Victor JD, Schiff ND (2012) Utilizing hierarchical decomposition to determine the causal nature of low frequency oscillations in unrecovered severe brain injury. Society for neuroscience 42nd annual meeting. Abstract 553.15
37. Chatelle C et al (2014) Changes in cerebral metabolism in patients with a minimally conscious state responding to zolpidem. Front Hum Neurosci 8:917
38. Liu AA, Voss HU, Dyke JP, Heier LA, Schiff ND (2011) Arterial spin labeling and altered cerebral blood flow patterns in the minimally conscious state. Neurology 77(16):1518–1523
39. Repucci MA, Schiff ND, Victor JD (2001) General strategy for hierarchical decomposition of multivariate time series: implications for temporal lobe seizures. Ann Biomed Eng 29: 1135–1149
40. Conte MM, Drover JD, Victor JD, Schiff ND (2012) Hierarchical decomposition (HD) of resting-state EEG in recovery following severe brain injury identifies causal influence of high frequency activity (beta band) over anterior forebrain. Society for neuroscience 42nd annual meeting. Abstract 553.16

Index

M.M. Monti, W.G. Sannita (eds.), *Brain Function and Responsiveness in Disorders of Consciousness*, DOI 10.1007/978-3-319-21425-2

MIX
Papier aus verantwortungsvollen Quellen
Paper from responsible sources
FSC® C105338
FSC
www.fsc.org

If you have any concerns about our products,
you can contact us on
ProductSafety@springernature.com

In case Publisher is established outside the EU,
the EU authorized representative is:
Springer Nature Customer Service Center GmbH
Europaplatz 3, 69115 Heidelberg, Germany

Printed by Libri Plureos GmbH
in Hamburg, Germany